Lecture Notes in Mathematics

Edited by A. Dold and B. Eckmann

768

Robert S. Doran
Josef Wichmann

Approximate Identities and Factorization in Banach Modules

Springer-Verlag
Berlin Heidelberg New York

Lecture Notes in Mathematics

For information about Vols. 1–551, please contact your bookseller or Springer-Verlag.

Vol. 552: C. G. Gibson, K. Wirthmüller, A. A. du Plessis and E. J. N. Looijenga. Topological Stability of Smooth Mappings. V, 155 pages. 1976.

Vol. 553: M. Petrich, Categories of Algebraic Systems. Vector and Projective Spaces, Semigroups, Rings and Lattices. VIII, 217 pages. 1976.

Vol. 554: J. D. H. Smith, Mal'cev Varieties. VIII, 158 pages. 1976.

Vol. 555: M. Ishida, The Genus Fields of Algebraic Number Fields. VII, 116 pages. 1976.

Vol. 556: Approximation Theory. Bonn 1976. Proceedings. Edited by R. Schaback and K. Scherer. VII, 466 pages. 1976.

Vol. 557: W. Iberkleid and T. Petrie, Smooth S^1 Manifolds. III, 163 pages. 1976.

Vol. 558: B. Weisfeiler, On Construction and Identification of Graphs. XIV, 237 pages. 1976.

Vol. 559: J.-P. Caubet, Le Mouvement Brownien Relativiste. IX, 212 pages. 1976.

Vol. 560: Combinatorial Mathematics, IV, Proceedings 1975. Edited by L. R. A. Casse and W. D. Wallis. VII, 249 pages. 1976.

Vol. 561: Function Theoretic Methods for Partial Differential Equations. Darmstadt 1976. Proceedings. Edited by V. E. Meister, N. Weck and W. L. Wendland. XVIII, 520 pages. 1976.

Vol. 562: R. W. Goodman, Nilpotent Lie Groups: Structure and Applications to Analysis. X, 210 pages. 1976.

Vol. 563: Séminaire de Théorie du Potentiel. Paris, No. 2. Proceedings 1975–1976. Edited by F. Hirsch and G. Mokobodzki. VI, 292 pages. 1976.

Vol. 564: Ordinary and Partial Differential Equations, Dundee 1976. Proceedings. Edited by W. N. Everitt and B. D. Sleeman. XVIII, 551 pages. 1976.

Vol. 565: Turbulence and Navier Stokes Equations. Proceedings 1975. Edited by R. Temam. IX, 194 pages. 1976.

Vol. 566: Empirical Distributions and Processes. Oberwolfach 1976. Proceedings. Edited by P. Gaenssler and P. Révész. VII, 146 pages. 1976.

Vol. 567: Séminaire Bourbaki vol. 1975/76. Exposés 471–488. IV, 303 pages. 1977.

Vol. 568: R. E. Gaines and J. L. Mawhin, Coincidence Degree, and Nonlinear Differential Equations. V, 262 pages. 1977.

Vol. 569: Cohomologie Etale SGA 4½. Séminaire de Géométrie Algébrique du Bois-Marie. Edité par P. Deligne. V, 312 pages. 1977.

Vol. 570: Differential Geometrical Methods in Mathematical Physics, Bonn 1975. Proceedings. Edited by K. Bleuler and A. Reetz. VIII, 576 pages. 1977.

Vol. 571: Constructive Theory of Functions of Several Variables, Oberwolfach 1976. Proceedings. Edited by W. Schempp and K. Zeller. VI. 290 pages. 1977

Vol. 572: Sparse Matrix Techniques, Copenhagen 1976. Edited by V. A. Barker. V, 184 pages. 1977.

Vol. 573: Group Theory, Canberra 1975. Proceedings. Edited by R. A. Bryce, J. Cossey and M. F. Newman. VII, 146 pages. 1977.

Vol. 574: J. Moldestad, Computations in Higher Types. IV, 203 pages. 1977.

Vol. 575: K-Theory and Operator Algebras, Athens, Georgia 1975. Edited by B. B. Morrel and I. M. Singer. VI, 191 pages. 1977.

Vol. 576: V. S. Varadarajan, Harmonic Analysis on Real Reductive Groups. VI, 521 pages. 1977.

Vol. 577: J. P. May, E_∞ Ring Spaces and E_∞ Ring Spectra. IV, 268 pages. 1977.

Vol. 578: Séminaire Pierre Lelong (Analyse) Année 1975/76. Edité par P. Lelong. VI, 327 pages. 1977.

Vol. 579: Combinatoire et Représentation du Groupe Symétrique, Strasbourg 1976. Proceedings 1976. Edité par D. Foata. IV, 339 pages. 1977.

Vol. 580: C. Castaing and M. Valadier, Convex Analysis and Measurable Multifunctions. VIII, 278 pages. 1977.

Vol. 581: Séminaire de Probabilités XI, Université de Strasbourg. Proceedings 1975/1976. Edité par C. Dellacherie, P. A. Meyer et M. Weil. VI, 574 pages. 1977.

Vol. 582: J. M. G. Fell, Induced Representations and Banach *-Algebraic Bundles. IV, 349 pages. 1977.

Vol. 583: W. Hirsch, C. C. Pugh and M. Shub, Invariant Manifolds. IV, 149 pages. 1977.

Vol. 584: C. Brezinski, Accélération de la Convergence en Analyse Numérique. IV, 313 pages. 1977.

Vol. 585: T. A. Springer, Invariant Theory. VI, 112 pages. 1977.

Vol. 586: Séminaire d'Algèbre Paul Dubreil, Paris 1975–1976 (29ème Année). Edited by M. P. Malliavin. VI, 188 pages. 1977.

Vol. 587: Non-Commutative Harmonic Analysis. Proceedings 1976. Edited by J. Carmona and M. Vergne. IV, 240 pages. 1977.

Vol. 588: P. Molino, Théorie des G-Structures: Le Problème d'Equivalence. VI, 163 pages. 1977.

Vol. 589: Cohomologie l-adique et Fonctions L. Séminaire de Géométrie Algébrique du Bois-Marie 1965–66, SGA 5. Edité par L. Illusie. XII, 484 pages. 1977.

Vol. 590: H. Matsumoto, Analyse Harmonique dans les Systèmes de Tits Bornologiques de Type Affine. IV, 219 pages. 1977.

Vol. 591: G. A. Anderson, Surgery with Coefficients. VIII, 157 pages. 1977.

Vol. 592: D. Voigt, Induzierte Darstellungen in der Theorie der endlichen, algebraischen Gruppen. V, 413 Seiten. 1977.

Vol. 593: K. Barbey and H. König, Abstract Analytic Function Theory and Hardy Algebras. VIII, 260 pages. 1977.

Vol. 594: Singular Perturbations and Boundary Layer Theory, Lyon 1976. Edited by C. M. Brauner, B. Gay, and J. Mathieu. VIII, 539 pages. 1977.

Vol. 595: W. Hazod, Stetige Faltungshalbgruppen von Wahrscheinlichkeitsmaßen und erzeugende Distributionen. XIII, 157 Seiten. 1977.

Vol. 596: K. Deimling, Ordinary Differential Equations in Banach Spaces. VI, 137 pages. 1977.

Vol. 597: Geometry and Topology, Rio de Janeiro, July 1976. Proceedings. Edited by J. Palis and M. do Carmo. VI, 866 pages. 1977.

Vol. 598: J. Hoffmann-Jørgensen, T. M. Liggett et J. Neveu, Ecole d'Eté de Probabilités de Saint-Flour VI – 1976. Edité par P.-L. Hennequin. XII, 447 pages. 1977.

Vol. 599: Complex Analysis, Kentucky 1976. Proceedings. Edited by J. D. Buckholtz and T. J. Suffridge. X, 159 pages. 1977.

Vol. 600: W. Stoll, Value Distribution on Parabolic Spaces. VIII, 216 pages. 1977.

Vol. 601: Modular Functions of one Variable V, Bonn 1976. Proceedings. Edited by J.-P. Serre and D. B. Zagier. VI, 294 pages. 1977.

Vol. 602: J. P. Brezin, Harmonic Analysis on Compact Solvmanifolds. VIII, 179 pages. 1977.

Vol. 603: B. Moishezon, Complex Surfaces and Connected Sums of Complex Projective Planes. IV, 234 pages. 1977.

Vol. 604: Banach Spaces of Analytic Functions, Kent, Ohio 1976. Proceedings. Edited by J. Baker, C. Cleaver and Joseph Diestel. VI, 141 pages. 1977.

Vol. 605: Sario et al., Classification Theory of Riemannian Manifolds. XX, 498 pages. 1977.

Vol. 606: Mathematical Aspects of Finite Element Methods. Proceedings 1975. Edited by I. Galligani and E. Magenes. VI, 362 pages. 1977.

Vol. 607: M. Métivier, Reelle und Vektorwertige Quasimartingale und die Theorie der Stochastischen Integration. X, 310 Seiten. 1977.

Vol. 608: Bigard et al., Groupes et Anneaux Réticulés. XIV, 334 pages. 1977.

continuation on page 309

Lecture Notes in Mathematics

Edited by A. Dold and B. Eckmann

768

Robert S. Doran
Josef Wichmann

Approximate Identities and Factorization in Banach Modules

Springer-Verlag
Berlin Heidelberg New York 1979

Authors

Robert S. Doran
Department of Mathematics
Texas Christian University
Fort Worth, Texas 76129
USA

Josef Wichmann
Department of Mathematical Sciences
University of Petroleum & Minerals
Dhahran, Saudi Arabia

AMS Subject Classifications (1970): 22B10, 43A20, 46H05, 46H25, 46L05, 46L20

ISBN 3-540-09725-2 Springer-Verlag Berlin Heidelberg New York
ISBN 0-387-09725-2 Springer-Verlag New York Heidelberg Berlin

Printing and binding: Beltz Offsetdruck, Hemsbach/Bergstr.
2141/3140-543210

TO

PAUL J. COHEN

AND

EDWIN HEWITT

Math

Sep

PREFACE

In recent years Banach algebras with an approximate identity have received an increasing amount of attention. Many results known for C^*-algebras or for algebras with an identity have been extended to algebras with an approximate identity. This is an important extension since it includes the group algebra $L^1(G)$ of a locally compact group G and many other naturally occuring algebras.

Surprisingly little is known about the approximate identities themselves. In this monograph we have tried to collect all basic results about them with the aim of stimulating further research in this direction. As the main tool in the study of Banach algebras with bounded approximate identity we present Cohen-Hewitt's factorization theorem for Banach modules and its many refinements. Recognizing that factorization theory is a subject of great importance in its own right, we have included the most recent and up-to-date results in this area that we had knowledge of.

The level of exposition should be appropriate for those who are familiar with basic real and complex analysis, the elementary theory of commutative Banach algebras, and first results concerning C^*-algebras (say, a nodding acquaintence with the first two chapters of Dixmier [70]). Granted these, the monograph contains complete proofs, although we should warn the reader that some of the arguments are a bit tedious and will require some diligence on his part. We have assumed throughout that all algebras are over the complex field. The interested reader can determine, by examining a given definition or theorem, if the complexes can be replaced by the reals.

Examples and counterexamples are discussed whenever we knew of them. There is much further research to be done, and we have indicated in the last chapter a number of unsolved problems. The reader will also find in this last chapter a description of many interesting results, with references, which could not be included in the text proper without greatly increasing the size of this volume. A comprehensive bibliography concerning approximate identities and factorization has been assembled with both the Mathematical Review number and Zentralblatt number attached to aid the reader in his study of the subject. Although a reasonable attempt has been made in the last chapter to cite appropriate sources, omissions have undoubtedly occured and we apologize in advance to those we may have overlooked.

The authors are deeply grateful to Robert B. Burckel and Barry E. Johnson for their contributions to the present volume. Dr. Burckel carefully read two earlier versions of the manuscript and suggested many corrections, additions, and improvements; to him we offer our deepest thanks. Dr. Johnson also read an earlier version of the manuscript and made many helpful suggestions which have been incorporated into this volume. In addition, we wish to thank Peter G. Dixon for valuable correspondence, preprints, and reprints. Much of what is in this volume concerning approximate identities is due to him. His willingness to communicate his work, even when very busy, has been a big help to us. We thank him too for pointing out the recent work of Allan M. Sinclair, and then we must thank Dr. Sinclair for graciously allowing us to include some of his beautiful results prior to their publication. Many other friends have sent us their work, and to them we are also grateful. We wish to thank Ronald L. Morgan, a student of the first author, for reading the final manuscript and pointing out a number of slips which had gone undetected. Of course, any remaining errors or inaccuracies are the sole responsibility of the authors.

We wish to acknowledge partial financial support from Texas Christian University during the writing of this volume, and finally, we wish to thank Shirley Doran for her meticulous typing of the entire manuscript.

Robert S. Doran

Josef Wichmann

CONTENTS

CHAPTER I. APPROXIMATE IDENTITIES IN NORMED ALGEBRAS

§1. Approximate identities 2
§2. One-sided and two-sided 5
§3. Renorming 17
§4. Banach algebras with sequential approximate identities . 23
§5. Banach algebras with nonsequential approximate identities 33
§6. Normed algebras with sequential approximate identities . 40
§7. Quotients 42
§8. Tensor products 44
§9. Approximate units 50
§10. Topological zero divisors 57
§11. Topologically nilpotent elements 60
§12. C^*-algebras 66
§13. Group algebras 81

CHAPTER II. FACTORIZATION IN BANACH MODULES

§14. Banach modules 90
§15. Essential Banach modules 91
§16. Factorization 93
§17. Multiple factorization 114
§18. Power factorization 122
§19. Factorization and semigroups 129
§20. Analytic factorization I 140
§21. Analytic factorization II 146
§22. Factorable Banach algebras without approximate units . . 161
§23. Nonfactorization 167
§24. Fréchet modules 170

§25. Essential Fréchet modules 172

§26. Factorization in Fréchet modules 174

CHAPTER III. MORE ABOUT APPROXIMATE IDENTITIES

§27. Local versus global 182

§28. Well-behaved approximate identities 184

§29. Quasicentral approximate identities 196

§30. Compact operators 203

§31. Abstract Segal algebras 210

§32. Sums of subspaces 221

§33. Weak approximate identities and Arens products 223

§34. A theorem on continuous bilinear mappings 226

§35. A majorization theorem for C^*-algebras 229

§36. Approximate diagonals 232

NOTES AND REMARKS 237

APPENDIX 266

BIBLIOGRAPHY 276

SUBJECT INDEX 296

SYMBOL INDEX 301

AUTHOR INDEX 304

CHAPTER I

APPROXIMATE IDENTITIES IN NORMED ALGEBRAS

Our purpose in this chapter is to set forth in a systematic way some of the main properties of approximate identities in a normed algebra. Section 1 contains definitions and a useful reformulation of the definition of approximate identity. In Section 2 the relationship between left, right, and two-sided approximate identities is examined. Sections 3, 4 and 5 center on the problem of finding an analogue for approximate identities of the classical result that every normed algebra with identity has an equivalent algebra norm in which the identity has norm one. Sections 7 and 8 give basic results on approximate identities in quotient algebras and the tensor product of two normed algebras.

The notion of bounded approximate unit is defined and studied in Section 9 and we show, among other things, that a normed algebra has a bounded approximate unit if and only if it has a bounded approximate identity. The relationships between topological divisors of zero, topologically nilpotent elements and approximate identities are studied in Sections 10 and 11. Approximate identities in C^*-algebras is the topic of Section 12. The main result is that every C^*-algebra contains an increasing approximate identity bounded by one. Finally we end the chapter with a brief look at approximate identities in the group algebra of a locally compact group.

I. APPROXIMATE IDENTITIES IN NORMED ALGEBRAS

1. Approximate identities.

(1.1) Definitions. Let A be a normed algebra. A net $\{e_\lambda\}_{\lambda\varepsilon\Lambda}$ in A is called a left (resp. right, two-sided) approximate identity, abbreviated l.a.i. (resp. r.a.i., t.a.i.), if for all $x \varepsilon A$,

$$\lim_{\lambda\varepsilon\Lambda} e_\lambda x = x,$$

(resp. $\lim\limits_{\lambda\varepsilon\Lambda} xe_\lambda = x$, $\lim\limits_{\lambda\varepsilon\Lambda} e_\lambda x = x = \lim xe_\lambda$). It is said to be bounded if there is a constant K such that $||e_\lambda|| \leq K$ for all $\lambda \varepsilon \Lambda$; in this case, we define the bound of $\{e_\lambda\}_{\lambda\varepsilon\Lambda}$ by $\sup \{||e_\lambda||: \lambda \varepsilon \Lambda\}$, and the norm by

$$||\{e_\lambda\}|| = \limsup_\lambda ||e_\lambda||.$$

An approximate identity $\{e_\lambda\}_{\lambda\varepsilon\Lambda}$ is sequential if Λ is the set of positive integers with the usual order, and is said to be countable if it has countable range. It is abelian (or commuting) if e_{λ_1} and e_{λ_2} commute for all $\lambda_1, \lambda_2 \varepsilon \Lambda$.

Remark. The notions of bound and norm of an approximate identity are closely related: if $\{e_\lambda\}$ is a l.a.i. of norm K, with $e_\lambda \neq 0$ for each λ, then $\{K||e_\lambda||^{-1}e_\lambda\}$ is a l.a.i. of bound K. The norm of an approximate identity is always ≥ 1. Indeed,

$$||x|| = \lim_\lambda ||e_\lambda x|| \leq \limsup_\lambda ||e_\lambda||\cdot||x|| \quad \text{for all } x \varepsilon A,$$

so $1 \leq \limsup\limits_\lambda ||e_\lambda||$.

§1. APPROXIMATE IDENTITIES

(1.2) Proposition. A normed algebra A has a l.a.i. (bounded by K) if and only if for every finite set $\{x_1,\ldots,x_n\}$ of elements in A and every $\varepsilon > 0$ there exists an element $e \in A$ (with $\|e\| \le K$) such that $\|x_i - ex_i\| < \varepsilon$ for $i = 1,\ldots,n$.

Proof. Let A be a normed algebra with a l.a.i. $\{e_\lambda\}_{\lambda\in\Lambda}$ (bounded by K). Consider a finite set $\{x_1,\ldots,x_n\}$ of elements in A and choose $\varepsilon > 0$. Since

$$\lim_{\lambda\in\Lambda} e_\lambda x_i = x_i \quad \text{for} \quad i = 1,\ldots,n$$

there exist $\lambda_i \in \Lambda$ $(i = 1,\ldots,n)$ such that for $i = 1,\ldots,n$

$$\|x_i - e_\lambda x_i\| < \varepsilon \quad \text{for all} \quad \lambda \ge \lambda_i.$$

Since Λ is a directed set there exists a $\lambda_o \in \Lambda$ with

$$\lambda_o \ge \lambda_i \quad \text{for all} \quad i = 1,\ldots,n.$$

Then for $e_{\lambda_o} \in A$ (with $\|e_{\lambda_o}\| \le K$) we have

$$\|x_i - e_{\lambda_o} x_i\| < \varepsilon \quad \text{for} \quad i = 1,\ldots,n.$$

Conversely, assume that A is a normed algebra with the property that (there exists a constant K such that) for every finite set $\{x_1,\ldots,x_n\}$ of elements in A and every $\varepsilon > 0$ there exists an element $e \in A$ (with $\|e\| \le K$) such that

$$\|x_i - ex_i\| < \varepsilon \quad \text{for all} \quad i = 1,\ldots,n.$$

Denote by Λ the set of all pairs $\lambda = (F,n)$ with F a finite subset of A and $n = 1,2,\ldots$; Λ is a directed set with respect to the partial ordering $\leq$ defined by: $(F_1,n_1) \leq (F_2,n_2)$ iff $F_1 \subset F_2$ and $n_1 \leq n_2$. Then for each $\lambda = (F,n)$ in Λ there is an element $e_\lambda \in A$ (with $||e_\lambda|| \leq K$) such that

$$||x - e_\lambda x|| < \frac{1}{n} \text{ for all } x \in F.$$

Thus for every $x \in A$ and $\varepsilon > 0$ there is a $\lambda_o \in \Lambda$ such that

$$||x - e_\lambda x|| < \varepsilon \text{ for all } \lambda \geq \lambda_o;$$

i.e., the net $\{e_\lambda\}_{\lambda\in\Lambda}$ (bounded by K) has the property:

$$\lim_{\lambda\in\Lambda} e_\lambda x = x \text{ for all } x \in A.$$

Hence $\{e_\lambda\}_{\lambda\in\Lambda}$ is a l.a.i. for A (bounded by K). □

(1.3) Proposition. A normed algebra A has a t.a.i. (bounded by K) if and only if for every finite set $\{x_1,\ldots,x_n\}$ of elements in A and every $\varepsilon > 0$ there exists an element $e \in A$ (with $||e|| \leq K$) such that $||x_i - ex_i|| < \varepsilon$ and $||x_i - x_ie|| < \varepsilon$ for all $i = 1,\ldots,n$.

Proof. Analogous to the proof of Proposition (1.2). □

The following elementary result will be very helpful.

(1.4) Lemma. Let D be a dense subset of a normed algebra A. If A has a l.a.i. (r.a.i., t.a.i.) $\{e_\lambda\}_{\lambda\in\Lambda}$ then A has a l.a.i. (r.a.i., t.a.i.) $\{f_\mu\}_{\mu\in M}$ such that $f_\mu \in D$ ($\mu \in M$). If, further, $\{e_\lambda\}_{\lambda\in\Lambda}$ is sequential or bounded by K, then $\{f_\mu\}_{\mu \varepsilon M}$ may be chosen to have the same property. Conversely, if there is in D a net $\{g_\nu\}$ which is a bounded l.a.i. (r.a.i., t.a.i.) for D, then $\{g_\nu\}$ is a bounded l.a.i. (r.a.i., t.a.i.) for A.

<u>Proof</u>. This is straightforward. □

(1.5) <u>Proposition</u>. Let A be a separable normed algebra. If A has a bounded l.a.i., then it has a sequential l.a.i. bounded by the same constant.

<u>Proof</u>. Let $\{x_n\}_{n=1}^{\infty}$ be a countable dense subset of A and let $\{e_\lambda\}_{\lambda \varepsilon \Lambda}$ be a bounded l.a.i. in A. For each $n = 1,2,\ldots$ choose $\lambda_n \varepsilon \Lambda$ such that $||x_i - e_{\lambda_n} x_i|| < \frac{1}{n}$ for all $i = 1,\ldots,n$. Then $\{e_{\lambda_n}\}_{n=1}^{\infty}$ is a sequential l.a.i. bounded by the same constant. □

2. <u>One-sided and two-sided</u>.

We shall now study the relationship between left, right, and two-sided approximate identities.

In general the existence of a l.a.i. and a r.a.i. in an algebra does not imply the existence of a t.a.i. Our counterexample will be the semigroup algebra of a semigroup which is itself defined by specifying generators and relations. To give an explicit description of the elements of the semigroup we shall need the following lemma.

(2.1) <u>Lemma</u>. Let S be the semigroup on a set of generators T with relations

(1) $t_1t_2 = \gamma(t_1,t_2) \in T \quad ((t_1,t_2) \in R \subset T \times T)$.

Then the following are equivalent:

i) whenever $(t_1,t_2) \in R$ and $(t_2,t_3) \in R$ we have

a) $(t_1,t_2t_3) \in R$ if and only if $(t_1t_2,t_3) \in R$; and

b) if $(t_1,t_2t_3) \in R$ and $(t_1t_2,t_3) \in R$, then $\gamma(t_1,t_2t_3) = \gamma(t_1t_2,t_3)$; and, if $(t_1,t_2t_3) \notin R$ and $(t_1t_2,t_3) \notin R$, then $\gamma(t_1,t_2) = t_1$ and $\gamma(t_2,t_3) = t_3$ (i.e., $t_1t_2 = t_1$ and $t_2t_3 = t_3$).

ii) the distinct elements of S are those words $t_1\ldots t_n$ $(t_1,\ldots,t_n \in T)$ with $(t_j,t_{j+1}) \notin R$ $(1 \leq j \leq n-1)$.

Proof. Clearly ii) implies i).

Conversely, assume i). Suppose that a word

$$w = t_1\ldots t_n$$

reduces in two different ways by single applications of (1):

$$w_1 = t_1\ldots(t_it_{i+1})\ldots t_n,$$

$$w_2 = t_1\ldots(t_jt_{j+1})\ldots t_n.$$

Then each of the words w_1 and w_2 can be reduced, by single applications of (1), to a word w_3. This step uses i) if $i = j \pm 1$. An easy inductive argument shows that the full reduction of w to an irreducible word is unique. □

Condition ii) says that the elements of S are those words which cannot be reduced in length by applying (1). It is clear that every element of S can be written in this form, but the crucial point of ii) is that different irreducible words give different elements of S. The lemma asserts that to obtain this, it is sufficient to check that every word of length three reduces uniquely.

(2.2) <u>Example.</u> There is a Banach algebra with an abelian sequential unbounded l.a.i., an abelian sequential unbounded r.a.i., and no t.a.i.

<u>Proof</u>. Let A_o be the complex associative algebra generated by $\{t_i: i = 1,2,3,\ldots\}$ subject to the relations

$$t_i t_j = t_{\min\{i,j\}},$$

unless i is odd and j is even.

Using Lemma (2.1), we find that every element of A_o is uniquely expressible in the form

$$x = \sum_{n=1}^{\infty} \alpha_n t_n + \sum_{i,j} \alpha_{ij} t_i t_j, \tag{1}$$

where almost all of the scalars α_n, α_{ij} are zero, and the second summation is over odd values of i and even values of j.

For the rest of this proof, h, i will always denote odd numbers and j, k even numbers.

We define a norm on A_o by

$$||x|| = \sum_{n=1}^{\infty} |\alpha_n| 2^n + \sum_{i,j} |\alpha_{ij}| 2^{i+j}, \tag{2}$$

and form the completion A. A typical element x in A is of the form (1), without the restriction that almost all the α_n, α_{ij} should vanish, but with $||x||$, as in (2), being finite (for a proof that the norm in (2) is submultiplicative, see Appendix.)

It is now easy to check that $\{t_i\}$ is an abelian sequential unbounded r.a.i. in A, and $\{t_j\}$ an abelian sequential unbounded l.a.i. in A.

Suppose A has a t.a.i. $\{e_\lambda\}_{\lambda\in\Lambda}$. Let

$$e_\lambda = \sum_{n=1}^{\infty} \alpha_n^{(\lambda)} t_n + \sum_{i,j} \alpha_{ij}^{(\lambda)} t_i t_j .$$

The coefficient of t_h in $t_h e_\lambda - t_h$ is $\sum_{i\geq h} \alpha_i^{(\lambda)} - 1.$

This must tend to zero as λ runs through Λ. So

$$\lim_{\lambda\in\Lambda} \sum_{i\geq h} \alpha_i^{(\lambda)} = 1. \tag{3}$$

Similarly, considering the sum of the $t_i t_j$ terms in $xe_\lambda - x$, where $x = \sum_k 2^{-2k} t_k$, we obtain

$$\lim_{\lambda\in\Lambda} \sum_{i,j} |\alpha_{ij}^{(\lambda)}| 2^{j-i} = 0. \tag{4}$$

Finally, consider

$$e_\lambda x - x = \left(\sum_{n=1}^{\infty} \alpha_n^{(\lambda)} t_n + \sum_{i,j} \alpha_{ij}^{(\lambda)} t_i t_j\right)\left(\sum_k 2^{-2k} t_k\right) - \sum_k 2^{-2k} t_k .$$

The sum of the $t_i t_{i+1}$ terms in $e_\lambda x - x$ is

$$\alpha_i^{(\lambda)} t_i 2^{-2(i+1)} t_{i+1} + \alpha_{ii+1}^{(\lambda)} t_i t_{i+1}\left(\sum_{k\geq i+1} 2^{-2k} t_k\right) + \sum_{j>i+1} \alpha_{ij}^{(\lambda)} t_i t_j 2^{-2(i+1)} t_{i+1}$$

$$= \alpha_i^{(\lambda)} 2^{-2(i+1)} t_i t_{i+1} + \alpha_{ii+1}^{(\lambda)} \left(\sum_{k \geq i+1} 2^{-2k} \right) t_i t_{i+1} + \sum_{j>i+1} \alpha_{ij}^{(\lambda)} 2^{-2(i+1)} t_i t_{i+1}$$

$$= 2^{-2(i+1)} \left[\alpha_i^{(\lambda)} + \alpha_{ii+1}^{(\lambda)} \left(\sum_{k \geq 0} 2^{-2k} \right) + \sum_{j>i+1} \alpha_{ij}^{(\lambda)} \right] t_i t_{i+1}$$

$$= 2^{-2(i+1)} \left[\alpha_i^{(\lambda)} + \frac{16}{15} \alpha_{ii+1}^{(\lambda)} + \sum_{j>i+1} \alpha_{ij}^{(\lambda)} \right] t_i t_{i+1}.$$

Hence

$$\lim_{\lambda \in \Lambda} \sum_i \left| \alpha_i^{(\lambda)} + \frac{1}{15} \alpha_{ii+1}^{(\lambda)} + \sum_{j>i} \alpha_{ij}^{(\lambda)} \right| = 0. \tag{5}$$

Then

$$\sum_i |\alpha_i^{(\lambda)}| \leq \sum_i \left| \alpha_i^{(\lambda)} + \frac{1}{15} \alpha_{ii+1}^{(\lambda)} + \sum_{j>i} \alpha_{ij}^{(\lambda)} \right| + \frac{1}{15} \sum_i |\alpha_{ii+1}^{(\lambda)}| + \sum_i \sum_{j>i} |\alpha_{ij}^{(\lambda)}|$$

$$\leq \sum_i \left| \alpha_i^{(\lambda)} + \frac{1}{15} \alpha_{ii+1}^{(\lambda)} + \sum_{j>i} \alpha_{ij}^{(\lambda)} \right| + \frac{16}{15} \sum_i \sum_{j>i} |\alpha_{ij}^{(\lambda)}|$$

$$\leq \sum_i \left| \alpha_i^{(\lambda)} + \frac{1}{15} \alpha_{ii+1}^{(\lambda)} + \sum_{j>i} \alpha_{ij}^{(\lambda)} \right| + \frac{16}{15} \sum_i \sum_{j>i} |\alpha_{ij}^{(\lambda)}| 2^{j-i}$$

which tends to zero, by (4) and (5); contradicting (3). Thus A has no t.a.i. □

(2.3) <u>Proposition</u>. Let A be a normed algebra. If A has a bounded l.a.i. and a r.a.i., then it has a t.a.i.

Proof. Let $\{e_\lambda\}$ be a l.a.i. bounded by K and $\{f_\mu\}$ be a r.a.i. of A. Consider a finite set $\{x_1,\dots,x_n\}$ of elements in A and choose $\varepsilon > 0$. Fix μ_o such that

$$||x_i - x_i(e_\lambda + f_{\mu_o} - f_{\mu_o}e_\lambda)|| \leq ||x_i - x_i f_{\mu_o}||(1 + K) < \varepsilon$$

for all λ and $i = 1,\dots,n$. Now fix λ_o such that

$$||x_i - (e_{\lambda_o} + f_{\mu_o} - f_{\mu_o}e_{\lambda_o})x_i|| \leq ||x_i - e_{\lambda_o}x_i||(1 + ||f_{\mu_o}||) < \varepsilon$$

for all $i = 1,\dots,n$. Thus $g = e_{\lambda_o} + f_{\mu_o} - f_{\mu_o}e_{\lambda_o}$ is an element in A such that

$$||x_i - x_i g|| < \varepsilon \quad \text{and} \quad ||x_i - gx_i|| < \varepsilon$$

for all $i = 1,\dots,n$; thus A has a t.a.i. □

The question arises as to whether a Banach algebra can have a bounded l.a.i. and an unbounded r.a.i. without also having a bounded r.a.i. The next two results show that this is, indeed, possible but only when the r.a.i. is not sequential.

(2.4) Example. There is a Banach algebra with a bounded l.a.i., an unbounded r.a.i., but no bounded r.a.i.

Proof. Let Q be an uncountable set, and let Λ be the directed set of all finite subsets of Q. Let S be the semigroup with generators s_λ, t_λ $(\lambda \in \Lambda)$ and relations as follows:

$$s_{\lambda_1} t_{\lambda_2} = \begin{cases} s_{\lambda_1}, & \text{if } \lambda_1 \leq \lambda_2, \\ t_{\lambda_2}, & \text{if } \lambda_1 > \lambda_2, \end{cases}$$

$$s_{\lambda_1} s_{\lambda_2} = \begin{cases} s_{\lambda_1}, & \text{if } \lambda_1 \leq \lambda_2, \\ s_{\lambda_2}, & \text{if } \lambda_1 > \lambda_2, \end{cases}$$

$$t_{\lambda_1} t_{\lambda_2} = \begin{cases} t_{\lambda_1}, & \text{if } \lambda_1 \leq \lambda_2, \\ t_{\lambda_2}, & \text{if } \lambda_1 > \lambda_2. \end{cases}$$

By Lemma (2.1), S consists of words $w = r_1 \cdots r_n$, where each r_i is either s_{λ_i} or t_{λ_i} and, for each i, either λ_i and λ_{i+1} are incomparable, or r_i is an t_{λ_i} and r_{i+1} is an $s_{\lambda_{i+1}}$. Let A_o be the semigroup algebra of S and A the completion of A_o with respect to the following norm

$$||\Sigma\, \alpha_w w|| = \Sigma\, |\alpha_w| \cdot ||w||,$$

where, if $w = r_1 \cdots r_n$,

$$||w|| = \begin{cases} 1, & \text{if } r_1 \text{ is an } s_\lambda, \\ 2^{|\lambda|}, & \text{if } r_1 \text{ is a } t_\lambda. \end{cases}$$

Here $|\lambda|$ denotes the cardinality of λ; remember that λ is a finite subset of Q. The verification that this norm is submultiplicative is given in the Appendix.

That $\{s_\lambda\}_{\lambda\varepsilon\Lambda}$ is a bounded l.a.i. is immediate, using Lemma (2.1).

That $\{t_\lambda\}_{\lambda\varepsilon\Lambda}$ is an unbounded r.a.i. is more difficult. Let $x = \sum_{w\varepsilon S'} \alpha_w w$ be an element in A where S' is a subset of S and choose $\varepsilon > 0$. Then there is a finite subset S'' of S' such that

$$\sum_{w\varepsilon S'\setminus S''} |\alpha_w|\cdot||w|| < \frac{\varepsilon}{2}.$$

Now there are only finitely many $\lambda \varepsilon \Lambda$ such that s_λ or t_λ occurs in some word in S''. So, if λ_o is the union of all such λ, then $\lambda_o \varepsilon \Lambda$. Assuming, as we may, that $\alpha_w \neq 0$ for all $w \varepsilon S'$, we must have S' countable. Therefore, if Q' is the union of all λ such that s_λ or t_λ occurs in some word in S', then Q' is countable. Choose $q \varepsilon Q \setminus Q'$ and let $\lambda_1 = \lambda_o \cup \{q\}$. Then, for every $w \varepsilon S''$, $wt_{\lambda_1} = w$. Moreover, for every $w \varepsilon S'$, wt_{λ_1} either reduces to w or fails to reduce at all, since $q \varepsilon \lambda_1$ implies $\lambda_1 \not\leq \lambda$, whenever s_λ or t_λ occurs in w. Therefore

$$||x - xt_{\lambda_1}|| \leq 2 \sum_{w\varepsilon S'\setminus S''} |\alpha_w|\cdot||w|| < \varepsilon.$$

We now show that A has no bounded r.a.i. Let $\{f_\mu\}$ be a r.a.i. for A. Then given $x \varepsilon A$ and $\varepsilon > 0$, there exists $y \varepsilon \{f_\mu\}$ such that $||x - xy|| < \varepsilon$. Let $x = t_{\lambda_o}$, and let a corresponding $y = \sum_{w\varepsilon S'} \alpha_w w$. Then the coefficient of t_{λ_o} in $x - xy$ is $1 - \sum_{w\varepsilon S''} \alpha_w$, where every $w \varepsilon S''$

has initial letter t_λ $(\lambda \geq \lambda_o)$. Thus

$$||x - xy|| \geq (1 - \sum_{w \varepsilon S''} |\alpha_w|)2^{|\lambda_o|}$$

and

$$(\sum_{w \varepsilon S''} |\alpha_w|)2^{|\lambda_o|} \leq \sum_{w \varepsilon S''} |\alpha_w| \cdot ||w|| \leq ||y||.$$

So, if we put $\varepsilon = 2^{|\lambda_o|-1}$ and require y to be such that $||x - xy|| < \varepsilon$, we have $\sum_{w \varepsilon S''} |\alpha_w| > \frac{1}{2}$, and so $2^{|\lambda_o|-1} < ||y||$. But λ_o can be taken arbitrarily large, and $y = y(\lambda_o) \ \varepsilon \ \{f_\mu\}$ for all λ_o, so $\{f_\mu\}$ is unbounded.□

(2.5) Proposition. Let A be a Banach algebra. If A has a bounded l.a.i., then every sequential r.a.i. is bounded.

Proof. Let $\{e_\lambda\}$ be a l.a.i. bounded by K and $\{f_n\}$ a sequential r.a.i. of A. Observe that the operators T_n on A defined by

$$T_n x = xf_n \quad (x \ \varepsilon \ A)$$

are pointwise convergent. Hence, by the uniform boundedness theorem, there is a constant M such that

$$||T_n|| \leq M \text{ for all } n.$$

Then

$$||f_n|| = \lim_\lambda ||T_n e_\lambda|| \leq ||T_n|| \cdot \sup_\lambda ||e_\lambda|| \leq M \cdot K$$

for all n.□

(2.6) Proposition. Let A be a normed algebra. If A has a bounded l.a.i. $\{e_\lambda\}_{\lambda\varepsilon\Lambda}$ and a bounded r.a.i. $\{f_\mu\}_{\mu\varepsilon M}$, then $\{e_\lambda + f_\mu - f_\mu e_\lambda\}_{(\lambda,\mu)\varepsilon\Lambda\times M}$ is a bounded t.a.i. of A.

Proof. Suppose $\{e_\lambda\}_{\lambda\varepsilon\Lambda}$ is bounded by K. Then, for all $x \varepsilon A$,

$$||x - x(e_\lambda + f_\mu - f_\mu e_\lambda)|| \leq ||x - xf_\mu|| + ||x - xf_\mu||\cdot||e_\lambda||$$

$$\leq (1 + K)||x - xf_\mu||,$$

so $\{e_\lambda + f_\mu - f_\mu e_\lambda\}_{(\lambda,\mu)\varepsilon\Lambda\times M}$ is a l.a.i. Likewise, it is a r.a.i. It is obviously bounded. □

The preceding proposition shows that if a normed algebra has a l.a.i. of norm H and a r.a.i. of norm K, then it has a t.a.i. of norm at most $H + K + HK$. The next example demonstrates that this is virtually all that can be said about the relationship of the norms of l.a.i.'s, r.a.i.'s, and t.a.i.'s.

(2.7) Example. Given $H, K, J \geq 1$, such that

$$\sup\{H,K\} \leq J \leq H + K + HK,$$

there exists a Banach algebra A such that the infima of the norms of all l.a.i.'s, all r.a.i.'s, and all t.a.i.'s in A are equal to H, K, and J, respectively.

<u>Proof</u>. Let S be the semigroup generated by $\{t_n : n = 1,2,3,\ldots\}$ with relations $t_i t_j = t_{\min\{i,j\}}$ unless

(a) $i \equiv 2$ and $j \equiv 0 \pmod 3$, or

(b) $i \equiv 2$ and $j \equiv 1 \pmod 3$ and $i > j$, or

(c) $i \equiv 1$ and $j \equiv 0 \pmod 3$ and $i < j$.

By Lemma (2.1), the elements of S are:

$t_n \quad (n = 1,2,3,\ldots)$,

$t_i t_j \quad (i \equiv 2,\ j \equiv 0$ or $i \equiv 2,\ j \equiv 1,\ i > j$ or $i \equiv 1,\ j \equiv 0,\ i < j)$,

$t_i t_j t_k \quad (i \equiv 2,\ j \equiv 1,\ k \equiv 0,\ i > j,\ j < k)$.

From now on we use i, i' to denote integers congruent to 0 (mod 3), j, j' for integers congruent to 1, and k, k' for integers congruent to 2. We define $||\cdot|| : S \to R^+$ by: $||t_i|| = H$, $||t_j|| = J$, $||t_k|| = K$,

$||t_k t_i|| = KH$, $||t_k t_j|| = KJ \ (k > j)$, $||t_j t_i|| = JH \ (j < i)$,

$||t_k t_j t_i|| = KJH \quad (k > j,\ j < i)$.

It is straightforward that $||xy|| \leq ||x||\cdot||y|| \ (x, y \in S)$. Let A be the completion of the algebra A_o obtained from the semigroup S (see (2.2)). Then A is a Banach algebra and $\{t_i\}$, (respectively, $\{t_j\}$, $\{t_k\}$) is a left (respectively, two-sided, right) a.i. in A of norm H (respectively, J, K).

Let

$$E^\alpha = \sum_i \lambda_i^\alpha t_i + \sum_j \lambda_j^\alpha e_j + \sum_k \lambda_k^\alpha t_k + \sum_i \sum_k \lambda_{ki}^\alpha t_k t_i + \sum_{k>j}\sum \lambda_{kj}^\alpha t_k t_j + $$
$$+ \sum_{j<i}\sum \lambda_{ji}^\alpha t_j t_i + \sum_{\substack{k>j \\ j<i}}\sum\sum \lambda_{kji}^\alpha t_k t_j t_i$$

be a bounded l.a.i. for A. By considering the coefficient of $t_{i'}$ in $E^\alpha t_{i'} - t_{i'}$, we see that

$$\sum_{i\geq i'} \lambda_i^\alpha + \sum_{j>i'} \lambda_j^\alpha + \sum\sum_{i>j>i'} \lambda_{ji}^\alpha \to 1. \tag{1}$$

Since $J \geq H$ we have

$$||E^\alpha|| \geq \Sigma\, |\lambda_i^\alpha| H + \Sigma\, |\lambda_j^\alpha| J + \Sigma\Sigma\, |\lambda_{ji}^\alpha| JH$$

$$\geq (\Sigma\, |\lambda_i^\alpha| + \Sigma\, |\lambda_j^\alpha| + \Sigma\Sigma\, |\lambda_{ji}^\alpha|)H,$$

so $\limsup ||E^\alpha|| \geq H$. This proves that the infimum of the norms of all l.a.i.'s in A is equal to H. Similarly, if E^α is a bounded r.a.i. we have

$$\sum_{k\geq k'} \lambda_k^\alpha + \sum_{j>k'} \lambda_j^\alpha + \sum\sum_{k>j>k'} \lambda_{kj}^\alpha \to 1 \tag{2}$$

and hence $\limsup ||E^\alpha|| \geq K$.

Now if E^α is a bounded t.a.i. we have both (1) and (2). Further, by considering the coefficient of $t_{k'}t_{i'}$ in $E^\alpha t_{k'}t_{i'} - t_{k'}t_{i'}$, where $i' = k'+1$, we have

$$\sum_{i\geq i'} \lambda_i^\alpha + \sum_{j>i'} \lambda_j^\alpha + \sum_{k\geq k'} \lambda_k^\alpha + \sum\sum_{k>j>i'} \lambda_{kj}^\alpha + \sum\sum_{i>j>i'} \lambda_{ji}^\alpha + $$
$$+ \sum\sum_{k,i\geq k'} \lambda_{ki}^\alpha + \sum\sum\sum_{k,i>j>i'} \lambda_{kji}^\alpha \to 1. \tag{3}$$

Subtracting (1) and (2) from (3) gives

$$\sum\sum_{k,i\geq k'} \lambda_{ki}^\alpha + \sum\sum\sum_{k,i>j>i'} \lambda_{kji}^\alpha - \sum_{j>i'} \lambda_j^\alpha \to -1. \tag{4}$$

Writing $L^\alpha = \sum_{j>i'} \lambda_j^\alpha$, it follows that, given $\varepsilon > 0$, we have, for all sufficiently large α,

$$\Sigma\Sigma\, |\lambda_{ki}^\alpha| + \Sigma\Sigma\Sigma\, |\lambda_{kji}^\alpha| \geq \left| \sum\sum_{k,i\geq k'} \lambda_{ki}^\alpha + \sum\sum\sum_{k,i>j>i'} \lambda_{kji}^\alpha \right|$$
$$> |L^\alpha - 1| - \varepsilon \quad \text{by (4)}, \tag{5}$$

where the unlabelled Σ's denote summation over all values of the suffixes for which the summands are defined. Likewise, we have

$$\Sigma\,|\lambda_i^\alpha| + \Sigma\,\Sigma\,|\lambda_{ji}^\alpha| > |1 - L^\alpha| - \varepsilon, \text{ from (1)}, \tag{6}$$

and

$$\Sigma\,|\lambda_k^\alpha| + \Sigma\,\Sigma\,|\lambda_{kj}^\alpha| > |1 - L^\alpha| - \varepsilon, \text{ from (2)}, \tag{7}$$

for all sufficiently large α. Then

$$\begin{aligned} ||E^\alpha|| \geq\ & \Sigma\,|\lambda_j^\alpha|J + (\Sigma\,|\lambda_i^\alpha| + \Sigma\,\Sigma\,|\lambda_{ji}^\alpha|)H + (\Sigma\,|\lambda_k^\alpha| + \Sigma\,\Sigma\,|\lambda_{kj}^\alpha|)K \\ & + (\Sigma\,\Sigma\,|\lambda_{ki}^\alpha| + \Sigma\,\Sigma\,\Sigma\,|\lambda_{kji}^\alpha|)HK, \quad \text{since } J \geq 1, \\ \geq\ & |L^\alpha|J + (|1 - L^\alpha| - \varepsilon)(H + K + HK), \text{ by (5), (6), (7)}, \\ \geq\ & (|L^\alpha| + |1 - L^\alpha| - \varepsilon)J, \qquad \text{since } H + K + HK \geq J, \\ \geq\ & (1 - \varepsilon)J. \end{aligned}$$

Since this holds for all $\varepsilon > 0$, we have $\limsup ||E^\alpha|| \geq J$. This proves that the infimum of the norms of all t.a.i.'s is equal to J. This completes the proof of (2.7). □

3. Renorming.

Example (2.7) shows that one cannot generally find an approximate identity of norm one in a given algebra. Any normed algebra with identity 1 such that $||1|| \neq 1$ is a much easier example, since in an algebra with identity every approximate identity tends to 1 and so has norm equal to $||1||$.

It is therefore necessary to consider the idea of renorming the algebra, and finding an analogue of the classical theorem that every normed algebra with identity has an equivalent algebra norm in which the identity has norm one. However, the next example shows that the obvious analogue fails even for commutative Banach algebras.

(3.1) Example. There is a commutative Banach algebra A with a sequential a.i. of norm $H \geq 1$, which is of norm at least H in any other algebra norm on A.

Proof. Let A_o be the commutative algebra with generators s_i, t_i $(i = 1,2,3,...)$ and relations

$$s_i s_j = s_{\min\{i,j\}} \quad (i \neq j),$$

$$s_i t_j = t_j \quad (i > j),$$

$$s_i t_i = Ht_i,$$

$$s_i t_j = 0 \quad (i < j),$$

$$t_i t_j = 0$$

for $i, j = 1,2,3,...$. It is easy to check that the elements of A_o are linear combinations of terms s_i^k (i, k = 1,2,3,...) and t_j (j = 1,2,3,...). Let A be the completion of A_o in the norm given by

$$\left\| \sum_{i,k} \alpha_{ik} s_i^k + \sum_j \alpha_j t_j \right\| = \sum_{i,k} |\alpha_{ik}| H^k + \sum_j |\alpha_j|.$$

Then $\{s_i\}$ is a sequential a.i. in A of norm H, and the equation $s_i t_i = Ht_i$ ensures that its norm is at least H in any other algebra norm on A. □

To find an approximate identity of norm one in a given normed algebra we are therefore forced to consider renorming the algebra and looking for another approximate identity. Here we meet with limited success.

(3.2) Lemma. Let S be a bounded multiplicative semigroup in a normed algebra $(A, ||\cdot||)$. Then there exists an equivalent algebra norm $||\cdot||'$ on A such that

$$||s||' \leq 1 \quad \text{for all } s \in S.$$

Proof. Note that, by embedding A into the algebra A_1 obtained by adjoining an identity to A, we may assume without loss of generality that $1 \in S$.

First take

$$||x||'' = \sup\{||sx|| : s \in S\} \quad (x \in A).$$

It is easily verified that $||\cdot||''$ is an algebra norm on A such that

$$||x|| \leq ||x||'' \leq M||x|| \quad (x \in A),$$

where

$$M = \sup\{||s||: s \in S\}.$$

Clearly

$$||sx||'' \leq ||x||'' \quad (s \in S,\ x \in A).$$

Now take

$$||a||' = \sup\{||ax||'': x \in A,\ ||x||'' \leq 1\} \quad (a \in A).$$

This gives an equivalent algebra norm on A such that

$$||s||' \leq 1 \quad \text{for all} \quad s \in S. \square$$

(3.3) <u>Theorem</u>. Let A be a commutative normed algebra. If A has a bounded sequential a.i., then there is an equivalent algebra norm on A for which there exists a sequential a.i. of norm one.

<u>Proof</u>. Let $\{e_n\}$ be a sequential a.i. of A, bounded by K, and let $\{\varepsilon_n\}$ be a sequence of positive real numbers, tending to zero. By choosing a subsequence of $\{e_n\}$, if necessary, we may assume that

$$||e_j - e_i e_j|| < \varepsilon_n \quad (i > j,\ \tfrac{1}{2}(n^2 - n + 2) \leq i \leq \tfrac{1}{2}(n^2 + n)).$$

Define

$$f_n = \frac{1}{n}(e_{\frac{1}{2}(n^2-n+2)} + e_{\frac{1}{2}(n^2-n+2)+1} + \ldots + e_{\frac{1}{2}(n^2+n)})$$

for $n = 1,2,3,\ldots$. Then $\{f_n\}$ is a sequential a.i. of A, bounded by K, such that

$$||f_j - f_i f_j|| < \varepsilon_i \quad (i > j).$$

We intend to show that there is an equivalent algebra norm on A with respect to which the a.i. $\{f_n\}$ has norm one. To apply the preceding lemma, we shall need to estimate the norms of products of the f_n.

We begin by estimating $||f_n^m||$:

$$||f_n^m|| \leq \frac{1}{n^m} \Sigma \, ||e_{i_1} \cdots e_{i_m}||,$$

where the sum is taken over all m-tuples $(i_1, \ldots, i_m)$ with

$$\tfrac{1}{2}(n^2 - n + 2) \leq i_j \leq \tfrac{1}{2}(n^2 + n) \quad (1 \leq j \leq m).$$

Now we prove by induction on m that

$$||e_{i_1} \cdots e_{i_m}|| \leq (1 + \varepsilon_n)^m K^r,$$

where r is the highest power of any single e_i occuring in the product. Clearly, this inequality holds for $m = 1$. For the general case, we consider two possibilities. First, if $r = m$, so that the product is just e_i^m; then the inequality clearly holds. Secondly, if $r \neq m$, then the product contains two different elements; say, e_i, e_j with $i > j$. Let us write a_o for $e_{i_1} \cdots e_{i_m}$, a_1 for a_o with one factor e_i omitted, and a_2 for a_1 with one factor e_j omitted. Then

$$a_o = a_1 + a_2(e_i e_j - e_j),$$

and so by the inductive hypothesis

$$||a_o|| \le ||a_1|| + ||a_2||\cdot||e_ie_j - e_j||$$

$$\le (1 + \varepsilon_n)^{m-1}K^r + (1 + \varepsilon_n)^{m-2}K^r\varepsilon_n$$

$$\le (1 + \varepsilon_n)^m K^r.$$

This proves our estimate for $||e_{i_1}\ldots e_{i_m}||$.

Observe that the number of such products with a given value of r is at most $n\binom{m}{r}n^{m-r}$, where $\binom{m}{r}$ denotes a binomial coefficient. The number of such products is also at most n^m, which is a better estimate when $r = 1$. Thus we obtain

$$||f_n^m|| \le n^{-m} \sum_{r=2}^{m} n\binom{m}{r}n^{m-r}(1 + \varepsilon_n)^m K^r + n^{-m}n^m(1 + \varepsilon_n)^m K$$

$$\le \sum_{r=0}^{m} \binom{m}{r}n^{-\frac{r}{2}} K^r(1 + \varepsilon_n)^m + K(1 + \varepsilon_n)^m$$

$$\le 2K(1 + \varepsilon_n)^m(1 + \frac{K}{\sqrt{n}})^m.$$

From this inequality we obtain by induction on $\sum_{j=1}^{s} m_j$ (in the same way that the estimate for $||e_{i_1}\ldots e_{i_m}||$ was established!) that

$$||f_{n_1}^{m_1}\ldots f_{n_s}^{m_s}|| \le 2K \prod_{j=1}^{s} \{(1 + \varepsilon_{n_j})^{m_j}(1 + \frac{K}{\sqrt{n_j}})^{m_j}\}$$

for all $n_1,\ldots,n_s$. Thus the elements

$$f_n/(1+\varepsilon_n)(1+\frac{K}{\sqrt{n}}) \quad (n = 1,2,3,\ldots),$$

form a bounded multiplicative semigroup in A. By the preceding lemma there exists an equivalent algebra norm $||\cdot||'$ on A such that

$$||f_n||' \leq (1+\varepsilon_n)(1+\frac{K}{\sqrt{n}}) \quad (n = 1,2,\ldots),$$

which implies that $\limsup ||f_n||' = 1$. This proves the theorem. □

Analyzing the proof of Theorem (3.3) we see that the following slightly more general result has been established:

(3.4) Theorem. If A is a normed algebra with an abelian sequential bounded l.a.i., then there is an equivalent norm in which A has an abelian sequential bounded l.a.i. of bound 1.

Our proof of the next theorem depends on the Cohen factorization method and will be given later (see (19.5)).

(3.5) Theorem. If A is a normed algebra with a sequential bounded t.a.i. $\{e_n\}$, then there is an equivalent norm in which A has a sequential bounded t.a.i. $\{f_n\}$ of bound 1. Further, if A is complete, then $\{f_n\}$ may be chosen abelian.

4. Banach algebras with sequential approximate identities.

The purpose of this section is to show that it is not possible to weaken the hypotheses of Theorems (3.4) or (3.5) to algebras which have only bounded left approximate identities. The examples constructed here and in the next two sections also appear to be of independent interest.

(4.1) Theorem. (i) For every $H > 1$ there exists a separable Banach algebra $(A,||.||)$ with a sequential bounded l.a.i. $\{e(n)\}$ of bound H such that for no equivalent norm $||.||'$ on A is there a bounded l.a.i. of $||.||'$-bound less than H.

(ii) For every $H \geq 1$ there exists a separable Banach algebra $(A,||.||)$ with a sequential bounded l.a.i. $\{e(n)\}$ such that:

a) for every $\varepsilon > 0$, there is an equivalent norm on A in which $\{e(n)\}$ has bound less than $H + \varepsilon$; and

b) for no equivalent norm $||.||'$ on A is there a bounded l.a.i. of $||.||'$-bound less than or equal to H.

Proof. Let $\underline{H} : Z^+ \to R^+$ be a function, to be specified later, which satisfies for all $m, n \in Z^+$:

i) $\underline{H}(n) \geq 1$;

ii) $\underline{H}(m + n) \leq \underline{H}(m)\underline{H}(n)$, and $\underline{H}$ is (weakly) monotonically increasing;

iii) for some $\rho > 0$, $\underline{H}(n) \geq \rho n$.

The construction of A is similar to the constructions used in earlier examples (e.g., see (2.2)). When we say that A is 'the Banach algebra with generators $e(n)$ $(n \in Z^+)$, relations $e(i)e(j) = e(j)$ $(i \geq j)$ and norms $||e(i_1)\ldots e(i_m)|| = \underline{H}(m)$ ', we mean that the following construction is being used: Let A_o be the algebra with generators and relations as stated. Using Lemma (2.1), we see that a typical element of A_o is of the form

$$x = \Sigma\, \lambda(i_1,\ldots,i_m)e(i_1)\ldots e(i_m), \tag{1}$$

§4. SEQUENTIAL APPROXIMATE IDENTITIES

where the summation is over a finite set of strictly increasing sequences $(i_1,\ldots,i_m)$ of positive integers, and the $\lambda(i_1,\ldots,i_m)$ are scalars. We define a norm on A_o by

$$||x|| = \Sigma\ |\lambda(i_1,\ldots,i_m)|\underline{H}(m).$$

Condition ii) on the function $\underline{H}$ ensures that this is an algebra norm. The Banach algebra A is then defined as the completion of A_o in this norm. A typical element of A is of the same form (1), except that the summation may be over an infinite set of sequences $(i_1,\ldots,i_m)$, subject to $\Sigma\ |\lambda(i_1,\ldots,i_m)|\ \underline{H}(m) < \infty$.

We shall need another norm on A. For a typical $x \in A$ as in (1), we define

$$||x||_1 = \Sigma\ |\lambda(i_1,\ldots,i_m)|.$$

Then $||x||_1 \leq ||x||$ for x in A, but the norms are not equivalent. With the same expression (1) for a typical $x \in A$, and with $E \subset Z^+$, we define

$$S(E,x) = \Sigma'\ \lambda(i_1,\ldots,i_m)e(i_1)\ldots e(i_m),$$

the summation Σ' being restricted to those sequences $(i_1,\ldots,i_m)$ such that $i_j \in E$ for some j; and

$$s(E,x) = \Sigma''\lambda(i_1,\ldots,i_m)e(i_1)\ldots e(i_m),$$

where the summation Σ'' is restricted to those sequences $(i_1,\ldots,i_m)$ such that $i_j \in E$ for all j. More generally, if $E_1,\ldots,E_n \subset Z^+$ we define

$$s(E_1,E_2,\ldots,E_n,x) = \Sigma'''\lambda(i_{11},\ldots,i_{nk_n})e(i_{11})\ldots e(i_{1k_1})e(i_{21})\ldots e(i_{nk_n}),$$

where the summation Σ''' is over those sequences $(i_{11},i_{12},\ldots,i_{1k_1},i_{21},$

$\ldots,i_{2k_2},i_{31},\ldots,i_{n1},\ldots,i_{nk_n})$ with $k_j \geq 1$ $(1 \leq j \leq n)$ and $i_{jp} \in E_j$ for $1 \leq j \leq n$, $1 \leq p \leq k_j$.

We assume that A has an algebra norm $||.||'$ equivalent to $||.||$, (suppose $||x|| \leq M\cdot||x||'$ for $x \in A$), and that A has a bounded l.a.i. of $||.||'$-bound K, and so of $||.||$-bound at most MK. For each $\varepsilon > 0$ and $n \in Z^+$, we select from this bounded l.a.i. an element $f_{n\varepsilon}$ such that

$$||f_{n\varepsilon}e(m) - e(m)|| < \varepsilon$$

for $m \leq n$.

(4.2) <u>Lemma</u>. For every $\varepsilon > 0$, there is a positive integer N such that, for every finite set C of integers greater than N,

$$||S(C,f_{n\varepsilon})||_1 < \varepsilon,$$

for arbitrarily large n.

<u>Proof</u>. If not, then there exists $\varepsilon > 0$ such that, for every positive integer N, there is a finite set $C(N)$ (which cannot be empty) of integers greater than N, and an integer $\nu(N)$ such that $||S(C(N),f_{n\varepsilon})||_1 \geq \varepsilon$ for all $n \geq \nu(N)$. Define, inductively, $N_1 = 1$, $N_{i+1} = \sup C(N_i)$. Then, for all $n \geq \max\{\nu(N_1),\ldots,\nu(N_k)\}$,

$$\sum_{i=1}^{k} ||S(C(N_i),f_{n\varepsilon})||_1 \geq \varepsilon k.$$

Now if $f_{n\varepsilon} = \Sigma\, \lambda_{n\varepsilon}(i_1,\ldots,i_m)e(i_1)\ldots e(i_m)$, then

$$\sum_{i=1}^{k} ||S(C(N_i),f_{n\varepsilon})||_1 = \Sigma\, q(i_1,\ldots,i_m)|\lambda_{n\varepsilon}(i_1,\ldots,i_m)|,$$

where the summation Σ is over all strictly increasing sequences $(i_1,\ldots,i_m)$ and, for each such sequence, the non-negative integer $q(i_1,\ldots,i_m)$ is the number of integers i for which $\{i_1,\ldots,i_m\} \cap C(N_i) \neq \emptyset$ and $1 \leq i \leq k$. Since the sets $C(N_i)$ are pairwise disjoint, $q(i_1,\ldots,i_m) \leq m$. Thus

$$\begin{aligned}\varepsilon k &\leq \Sigma\ m|\lambda_{n\varepsilon}(i_1,\ldots,i_m)| \\ &\leq \rho^{-1}\ \Sigma\ \underline{H}(m)\ |\lambda_{n\varepsilon}(i_1,\ldots,i_m)| \\ &= \rho^{-1}\ ||f_{n\varepsilon}|| \\ &\leq \rho^{-1}\ MK.\end{aligned}$$

Letting $k \to \infty$, we obtain a contradiction, which proves the lemma [the only property of the sequence $\{f_{n\varepsilon}\}$ which we have used is that it is $||.||$-bounded; but this extra generality will not be needed].□

<u>Proceeding with the proof of</u> (4.1), we fix $\varepsilon > 0$ arbitrarily. This gives us a positive integer N for which the conclusion of the lemma holds. We define an increasing sequence of positive integers c_n $(n = 0,1,2,\ldots)$ and a sequence f_n $(n = 1,2,3,\ldots)$ selected from $\{f_{m\varepsilon} : m \in Z^+\}$. Let $c_o = N$. Having defined $c_o, c_1,\ldots,c_{n-1}$, we apply the lemma with $C = (N,c_{n-1}]$ $(= \{N+1,N+2,\ldots,c_{n-1}\})$ and find $m > c_{n-1}$ such that

$$||S(C,f_{m\varepsilon})||_1 < \varepsilon. \tag{2}$$

We define f_n to be this $f_{m\varepsilon}$. We then choose $c_n > c_{n-1}$ such that

$$||S((c_n,\infty),f_n)||_1 < \varepsilon. \tag{3}$$

We write $f_n = u_n + v_n + w_n$, where

$$u_n = s((c_{n-1},c_n],f_n),$$

$$v_n = S([1,N],s([1,N] \cup (c_{n-1},c_n],f_n)),$$

$$w_n = S((N,c_{n-1}],f_n) + S((c_n,\infty),f_n).$$

By (2) and (3), $||w_n||_1 < 2\varepsilon$.

We shall consider the product $f_1 f_2 \ldots f_n$. From this, we shall isolate the part $u_1 u_2 \ldots u_n$; a product in which no reduction occurs. We shall find a lower estimate for $||u_1 u_2 \ldots u_n||$ and hence a lower estimate for $||f_1 f_2 \ldots f_n||$. Doing this for all n, we shall obtain a lower estimate for $\sup \{||f_n||' : n \in Z^+\}$. First, let

$$t_n(x) = s((c_o,c_1],(c_1,c_2],\ldots,(c_{n-1},c_n],x)$$

for $x \in A$. Notice that, since each monomial of $t_n(x)$ has length at least n,

$$||t_n(x)|| \geq \underline{H}(n)||t_n(x)||_1 \qquad (x \in A). \tag{4}$$

Now

$$||t_n(f_1 \ldots f_n) - t_n((u_1+v_1)\ldots(u_n+v_n))||_1$$

$$= ||t_n(\sum_{j=1}^{n}(u_1+v_1)\ldots(u_{j-1}+v_{j-1})w_j f_{j+1}\ldots f_n)||_1$$

$$\leq \sum_{j=1}^{n} ||u_1+v_1||_1 \ldots ||u_{j-1}+v_{j-1}||_1 \cdot ||w_j||_1 \cdot ||f_{j+1}||_1 \ldots ||f_n||_1$$

$$\leq \sum_{j=1}^{n} ||f_1||_1 \ldots ||f_{j-1}||_1 \cdot ||w_j||_1 \cdot ||f_{j+1}||_1 \ldots ||f_n||_1$$

$$\leq \sum_{j=1}^{n} ||f_1|| \ldots ||f_{j-1}|| \cdot ||w_j||_1 \cdot ||f_{j+1}|| \ldots ||f_n||$$

$$\leq \ 2\varepsilon n(MK)^{n-1}, \tag{5}$$

where we have used the linearity of t_n, the inequalities $||t_n(x)||_1 \leq ||x||_1$ $(x \in A)$ and $||u_j + v_j||_1 \leq ||f_j||_1$ $(1 \leq j \leq n)$ (both of which hold because the deletion of one or more monomials from an element of A can only reduce its $||.||_1$-norm), and the fact that $||x||_1 \leq ||x||$ for all $x \in A$.

Consider $t_n((u_1+v_1)\ldots(u_n+v_n))$. Every monomial of every v_j contains some $e(i)$ with $1 \leq i \leq N$. But the defining relations for the multiplication on A imply that a product $e(i)e(j)$ is either left as $e(i)e(j)$ or reduced to $e(\min\{i,j\})$. Thus every product $u_1\ldots v_j\ldots u_n$ involving one or more v's will still, after reduction, contain some $e(i)$ with $1 \leq i \leq N$ in each monomial. Therefore,

$$t_n((u_1+v_1)\ldots(u_n+v_n)) = t_n(u_1u_2\ldots u_n). \tag{6}$$

Now each u_j involves only those $e(i)$ with $c_{j-1} < i \leq c_j$. Thus, when the product $u_1u_2\ldots u_n$ is formed, no reduction takes place. Hence

$$||t_n(u_1u_2\ldots u_n)||_1 = ||u_1u_2\ldots u_n||_1 = ||u_1||_1.||u_2||_1\ldots||u_n||_1. \tag{7}$$

By the definition of f_n, we have

$$||f_ne(c_{n-1}) - e(c_{n-1})||_1 < \varepsilon. \tag{8}$$

Now $f_ne(c_{n-1}) = u_ne(c_{n-1}) + v_ne(c_{n-1}) + w_ne(c_{n-1})$. Consider each summand in turn. We have $u_ne(c_{n-1}) = \alpha e(c_{n-1})$, where the scalar α is the sum of the coefficients of the monomials of u_n, and so $|\alpha| \leq ||u_n||_1$. The product $v_ne(c_{n-1})$ must have, in each monomial, some $e(i)$ with $1 \leq i \leq N$, (in symbols: $S([1,N], v_n e(c_{n-1})) = v_n e(c_{n-1})$), since the same is true of v_n. Consequently, $v_n e(c_{n-1})$ has no monomial of the form $\lambda e(c_{n-1})$, λ a scalar. Finally, $||w_n e(c_{n-1})||_1 \leq ||w_n||_1 < 2\varepsilon$. By (8), the coefficient of the monomial $e(c_{n-1})$ in $f_n e(c_{n-1})$ is greater than $1 - \varepsilon$, and so, by looking at this coefficient, we obtain

$$||u_n||_1 > 1 - 3\varepsilon. \tag{9}$$

Finally, we have

$$\begin{aligned}
MK^n &\geq M\cdot||f_1 f_2 \cdots f_n||' \\
&\geq ||f_1 f_2 \cdots f_n|| \\
&\geq ||t_n(f_1 f_2 \cdots f_n)|| \\
&\geq \underline{H}(n)||t_n(f_1 f_2 \cdots f_n)||_1, && \text{by (4),} \\
&\geq \underline{H}(n)[||t_n((u_1+v_1)\cdots(u_n+v_n))||_1 - 2\varepsilon n(MK)^{n-1}], && \text{by (5),} \\
&= \underline{H}(n)[||u_1||_1\cdot||u_2||_1\cdots||u_n||_1 - 2\varepsilon n(MK)^{n-1}], && \text{by (6) and (7),} \\
&> \underline{H}(n)[(1 - 3\varepsilon)^n - 2\varepsilon(MK)^{n-1}], && \text{by (9).}
\end{aligned}$$

Now ε was fixed arbitrarily, so

$$MK^n \geq \underline{H}(n) \quad (n \in Z^+).$$

Hence $K \geq \limsup \underline{H}(n)^{1/n}$. Choosing $\underline{H}(n) = H^n$ $(n \in Z^+)$ proves part (i) of (4.1), since A has a bounded l.a.i. $\{e(n)\}$ of bound H, but we have shown that, for any other bounded l.a.i. with bound K in a renorming $||.||'$, $K \geq H$.

§4. SEQUENTIAL APPROXIMATE IDENTITIES

To prove part (ii), we put $\underline{H}(1) = 2H$, $\underline{H}(n) = nH^n$ $(n \geq 2)$. (The eccentric definition of $\underline{H}(1)$ is needed to satisfy $\underline{H}(m+n) \leq \underline{H}(m)\underline{H}(n)$ for $m, n \in Z^+$.) Given $\varepsilon > 0$, choose a minimal $N \in Z^+$ subject to $N \geq 2$ and $(H+\varepsilon)^N > NH^N$; we also suppose $\varepsilon < (\sqrt{2}-1)H$. Define a function $\underline{H}'$ by

$$\underline{H}'(n) = \begin{cases} (H+\varepsilon)^n, & \text{if } 1 \leq n < N \\ nH^n, & \text{if } n > N . \end{cases}$$

Then (see the Appendix) $\underline{H}'$ is monotone, and $\underline{H}'(m+n) \leq \underline{H}'(m)\underline{H}'(n)$. The norm on A obtained by using $\underline{H}'$ in place of the function $\underline{H}$ is equivalent to the original norm, and it gives the bounded l.a.i. $\{e(n)\}$ a bound of $H+\varepsilon$. Now if, in some renorming, a bounded l.a.i. has bound K, then we have shown that $MK^n \geq \underline{H}(n) = nH^n$ $(n \geq N)$. Taking $n > M$, we see that $K > H$. This completes the proof of Theorem (4.1).□

(4.3) Corollary. The Banach algebras of Theorem (4.1) have bounded left approximate identities, but no abelian bounded left approximate identities.

Proof. The corollary is immediate from Theorem (3.4).□

This corollary still leaves open the possibility that a Banach algebra with bounded l.a.i. of bound one necessarily has an abelian bounded l.a.i.. To provide a counterexample, we adapt the example of (4.1) by putting $\underline{H}(n) = 1$ for all $n \in Z^+$. Notice that Lemma (4.2) now fails, e.g., if $f_{n\varepsilon} = e(1)e(2)\ldots e(n) - e(1)e(2)\ldots e(n-1) + e(n)$, so a different approach is needed.

(4.4) Theorem. There is a separable Banach algebra with a sequential bounded l.a.i. of bound 1 which has no abelian bounded l.a.i..

Proof. Our algebra is constructed as in (4.1), with $\underline{H}(n) = 1$ for all $n \in Z^+$. Suppose $\{f_\lambda\}_{\lambda\in\Lambda}$ is a commuting bounded l.a.i. with $||f_\lambda|| \leq K$ $(\lambda \in \Lambda)$. We define, by induction, an increasing sequence of positive integers c_n $(n = 0,1,2,3,\ldots)$ and a sequence f_n $(n = 1,2,3,\ldots)$ selected from $\{f_\lambda : \lambda \in \Lambda\}$. Let $c_o = 1$. Having defined $c_o, c_1, \ldots, c_{n-1}$ and $f_1, f_2, \ldots, f_{n-1}$, we select f_n so that

$$||f_n e(c_{n-1}) - e(c_{n-1})|| < \frac{1}{4} \tag{10}$$

and

$$||f_n f_i - f_i|| < \frac{1}{4} \quad (1 \leq i \leq n-1), \tag{11}$$

the latter condition being void if $n = 1$. We then choose $c_n > c_{n-1}$ such that

$$||S((c_n,\infty),f_n)|| < \frac{1}{4}. \tag{12}$$

We write $f_n = u_n + v_n + w_n$, where

$$u_n = s([c_{n-1},c_n],f_n),$$

$$v_n = S([1,c_{n-1}),s([1,c_n],f_n)),$$

$$w_n = S([c_n,\infty),f_n).$$

As before, we consider the coefficient of the monomial $e(c_{n-1})$ in the product $f_n e(c_{n-1})$. We have:

$$u_n e(c_{n-1}) = \alpha e(c_{n-1}), \quad \text{where} \quad |\alpha| \leq ||u_n||;$$

$$s(\{c_{n-1}\}, v_n e(c_{n-1})) = 0; \quad \text{and}$$

$$||w_n e(c_{n-1})|| < \frac{1}{4}, \qquad \text{by (12).}$$

From these and (10) we obtain

$$||u_n|| > \frac{1}{2}. \tag{13}$$

If $1 \leq i \leq j$ then, by hypothesis, $f_i f_j = f_j f_i$ and so (11) yields:

$$||s([c_{i-1}, c_i), f_i f_j) - u_i|| < \frac{1}{4}. \tag{14}$$

But

$$s([c_{i-1}, c_i), f_i f_j) = s([c_{i-1}, c_i), (u_i + w_i) s([c_{i-1}, c_i), f_j)). \tag{15}$$

So

$$\begin{aligned} \frac{1}{4} &< ||s([c_{i-1}, c_i), f_i f_j)||, && \text{by (13) and (14),} \\ &\leq ||u_i + w_i|| \cdot ||s([c_{i-1}, c_i), f_j)||, && \text{by (15)} \\ &\leq K \cdot ||s([c_{i-1}, c_i), f_j)||. \end{aligned}$$

Finally,

$$K \geq ||f_j|| \geq \sum_{i=1}^{j-1} ||s([c_{i-1}, c_i), f_j)|| \geq \frac{j-1}{4K}.$$

This, for all $j \geq 1$, gives the desired contradiction, and completes the proof of the theorem. □

5. Banach algebras with non-sequential approximate identities.

I. APPROXIMATE IDENTITIES IN NORMED ALGEBRAS

In this section we show that Theorems (3.4) and (3.5) can fail dramatically in Banach algebras which have only non-sequential approximate identities.

(5.1) Theorem. (i) For every $H > 1$ there exists a commutative Banach algebra $(A, ||.||)$ with a bounded a.i. of bound H such that for no equivalent norm $||\cdot||'$ on A is there a bounded a.i. of $||.||'$-bound less than H.

(ii) For every $H \geq 1$ there exists a commutative Banach algebra $(A, ||.||)$ with a bounded a.i. $\{e(\lambda)\}_{\lambda \varepsilon \Lambda}$ such that:

(a) for every $\varepsilon > 0$, there is an equivalent norm on A in which $\{e(\lambda)\}_{\lambda \varepsilon \Lambda}$ has bound less than $H + \varepsilon$; and

(b) for no equivalent norm $||.||'$ on A is there a bounded a.i. of $||.||'$-bound less than or equal to H.

Proof. Let $\underline{H} : Z^+ \to R^+$ be a weakly monotone function to be specified later, such that $\underline{H}(n) \geq 1$ $(n \varepsilon Z^+)$ and $\underline{H}(m + n) \leq \underline{H}(m)\underline{H}(n)$ $(m, n \varepsilon Z^+)$. Using a common set-theoretic convention, we write ω_2 for the smallest ordinal of cardinality $\aleph_2$ and for the set of all ordinals less than ω_2. Let Λ be the set of all finite non-empty subsets of ω_2, partially ordered by inclusion. Our algebra A will be the commutative Banach algebra with generators $e(\lambda)$ $(\lambda \varepsilon \Lambda)$, relations $e(\lambda)e(\mu) = e(\lambda)$ $(\lambda \subset \mu)$ and norms $||e(\lambda_1)\ldots e(\lambda_n)|| = \underline{H}(n)$. A typical element of A may be written

$$x = \Sigma\ \xi(\{\lambda_1,\ldots,\lambda_n\})e(\lambda_1)\ldots e(\lambda_n), \tag{16}$$

where the summation is over all sets $\{\lambda_1,\ldots,\lambda_n\}$ with no $\lambda_i \subset \lambda_j$ $(i \neq j)$, and the $\xi(\{\lambda_1,\ldots,\lambda_n\})$ are scalars such that

$$||x|| = \Sigma\ |\xi(\{\lambda_1,\ldots,\lambda_n\})|\ \underline{H}(n) < \infty.$$

We shall say that an ordinal $\alpha < \omega_2$ <u>occurs in</u> x if α belongs to λ_i for some λ_i in a set $\{\lambda_1,\ldots,\lambda_n\}$ with $\xi(\{\lambda_1,\ldots,\lambda_n\}) \neq 0$. We shall write $\Theta(x)$ for the set of ordinals occurring in x. Notice that only countably many ordinals can occur in any one element of A.

Assume that A has an algebra norm $||.||'$ equivalent to $||.||$ (suppose $||x|| \leq M\cdot||x||'$ for $x \in A$), and a bounded a.i. of $||.||'$-bound K, and so of $||.||$-bound at most MK. For each $\varepsilon > 0$ and $\alpha < \omega_2$, we select from this bounded a.i. an element $f_{\alpha\varepsilon}$ such that

$$||f_{\alpha\varepsilon}e(\{\alpha\}) - e(\{\alpha\})|| < \varepsilon. \qquad (17)$$

Note that α is in $\Theta(f_{\alpha\varepsilon})$ when $\varepsilon < 1$ (by considering the coefficient of the monomial $e(\{\alpha\})$ in the product $f_{\alpha\varepsilon}e(\{\alpha\})$).

(5.2) <u>Lemma</u>. For every $\varepsilon > 0$, there is a countable set $E \subset \omega_2$ such that, for every countable set $C \subset \omega_2$ with $C \cap E = \emptyset$,

$$C \cap \Theta(f_{\alpha\varepsilon}) = \emptyset$$

for arbitrarily large α.

The reader should compare this lemma with Lemma (4.2), to see the analogy between the proofs of (5.1) and (4.1).

<u>Proof</u>. Suppose the lemma is false. Then there exists $\varepsilon > 0$ such that, for every countable $E \subset \omega_2$, there is a countable $C(E) \subset \omega_2$ with

$C(E) \cap E = \emptyset$, and an ordinal $\gamma(E) < \omega_2$ such that $C(E) \cap \Theta(f_{\gamma\varepsilon}) \neq \emptyset$ for all $\gamma \geq \gamma(E)$. We define disjoint countable sets C_α and ordinals γ_α $(0 \leq \alpha < \omega_1)$, by transfinite induction. Let $C_o = \emptyset$, $\gamma_o = 0$. Having defined C_β, γ_β for all $\beta < \alpha$, we apply our hypothesis that the lemma is false, with

$$E = \bigcup_{\beta<\alpha} C_\beta,$$

and define C_α, γ_α to be the resulting $C(E)$, $\gamma(E)$. Thus, for all $\gamma \geq \gamma_\alpha$, $C_\alpha \cap \Theta(f_{\gamma\varepsilon}) \neq \emptyset$. Now, since $\gamma_\alpha < \omega_2$ for all $\alpha < \omega_1$, we have

$$\sup\{\gamma_\alpha : \alpha < \omega_1\} < \omega_2.$$

Let $\delta = \sup\{\gamma_\alpha : \alpha < \omega_1\}$. Then $\Theta(f_{\delta\varepsilon})$ intersects every C_α $(\alpha < \omega_1)$. Since the C_α are pairwise disjoint, it follows that $\Theta(f_{\delta\varepsilon})$ is uncountable, which is impossible. This proves the lemma. [Notice that we have not even used the fact that $\{f_{\alpha\varepsilon} : \varepsilon > 0, 0 \leq \alpha < \omega_2\}$ is bounded.] □

<u>We proceed with the proof of the theorem</u>. As in the proof of (4.1), we fix ε with $0 < \varepsilon < 1$ arbitrarily. This gives us a countable $E \subset \omega_2$ for which the conclusion of Lemma (5.2) holds. We define an increasing sequence of ordinals α_n $(n = 1,2,3,\ldots)$, as follows. Choose α_1 in $\omega_2 \setminus E$ arbitrarily. Having chosen $\alpha_1,\ldots,\alpha_{n-1}$, we define

$$C = [\Theta(f_{\alpha_1\varepsilon}) \cup \ldots \cup \Theta(f_{\alpha_{n-1}\varepsilon})] \setminus E$$

and then choose α_n in $\omega_2 \setminus E$ such that $C \cap \Theta(f_{\alpha_n\varepsilon}) = \emptyset$. Such a choice is possible by Lemma (5.2). We write f_i for $f_{\alpha_i\varepsilon}$. Then α_i belongs to $\Theta(f_i) \setminus E$, so the definition of α_i shows that $\alpha_i \notin \Theta(f_j)$ $(j < i)$, and the definition of α_j shows that $\alpha_i \notin \Theta(f_j)$ $(j > i)$.

§5. NONSEQUENTIAL APPROXIMATE IDENTITIES

Using a notation slightly different from that employed previously, we define, for $x \in A$ as in (16) and ordinals $\beta_1,\ldots,\beta_n < \omega_2$,

$$s(\beta_1,\ldots,\beta_n,x) = \Sigma' \xi(\{\lambda_{11},\ldots,\lambda_{nk_n}\})e(\lambda_{11})\ldots e(\lambda_{nk_n}),$$

where the summation Σ' is over those sets which may be labelled

$$\{\lambda_{11},\lambda_{12},\ldots,\lambda_{1k_1},\lambda_{21},\ldots,\lambda_{n1},\lambda_{n2},\ldots,\lambda_{nk_n}\}$$

with $k_i \geq 1$ $(1 \leq i \leq n)$ and $\beta_i \in \lambda_{ij}$ $(1 \leq i \leq n,\ 1 \leq j \leq k_i)$. We let $f_i = u_i + v_i$, where $u_i = s(\alpha_i,f_i)$, and write $t_n(x) = s(\alpha_1,\ldots,\alpha_n,x)$ for $x \in A$. Note that, since each monomial of $t_n(x)$ has length at least n,

$$||t_n(x)|| \geq \underline{H}(n)||t_n(x)||_1 \quad (x \in A), \tag{18}$$

where, as before, the notation $||.||_1$ means $||.||$ with the function $\underline{H}(n) = 1$ $(n \in Z^+)$.

We now discuss the product $f_1f_2\ldots f_n$. Consider a product $u_1\ldots v_i\ldots u_n$ containing one or more v's. By definition, every nonzero monomial of v_i involves some $e(\lambda)$ with $\alpha_i \notin \lambda$. But since $\alpha_j \notin \Theta(f_i)$ $(j \neq i)$, $\{\alpha_1,\ldots,\alpha_n\} \cap \lambda = \emptyset$. Now every monomial of the product must, after the necessary reductions, contain an $e(\mu)$ with μ a subset of some such λ. Hence $t_n(u_1\ldots v_i\ldots u_n) = 0$. So

$$t_n(f_1f_2\ldots f_n) = t_n(u_1u_2\ldots u_n). \tag{19}$$

Moreover, since $\alpha_i \in \lambda$ for every $e(\lambda)$ in every monomial of u_i and α_i occurs nowhere in u_j $(j \neq i)$, there is no reduction or cancellation when the product $u_1u_2\ldots u_n$ is formed. Therefore

$$||t_n(u_1u_2\ldots u_n)||_1 = ||u_1u_2\ldots u_n||_1 = ||u_1||_1\cdot||u_2||_1\ldots||u_n||_1. \qquad (20)$$

By (17), $||f_ie(\{\alpha_i\}) - e(\{\alpha_i\})||_1 < \varepsilon$. Now $v_ie(\{\alpha_i\})$ has no monomial of the form $\xi e(\{\alpha_i\})$, ξ a scalar; whereas $u_ie(\{\alpha_i\}) = \eta e(\{\alpha_i\})$, where η is a scalar with $|\eta| \leq ||u_i||_1$. Hence, by considering the coefficient of $e(\{\alpha_i\})$ in $f_ie(\{\alpha_i\})$, we obtain

$$||u_i||_1 \;>\; 1 - \varepsilon. \qquad (21)$$

Finally, we have

$$\begin{aligned} MK^n &\geq M\cdot||f_1f_2\ldots f_n||' \\ &\geq ||f_1f_2\ldots f_n|| \\ &\geq ||t_n(f_1f_2\ldots f_n)|| \\ &\geq \underline{H}(n)||t_n(f_1f_2\ldots f_n)||_1, \quad \text{by (18)}, \\ &> \underline{H}(n)(1-\varepsilon)^n, \quad \text{by (19), (20) and (21).} \end{aligned}$$

So, since ε can be arbitrarily small, $MK^n \geq \underline{H}(n)$ $(n \in Z^+)$. The result then follows as in the proof of (4.1).□

(5.3) Theorem. There is a Banach algebra with a (non-sequential) bounded t.a.i. of bound 1 and no abelian l.a.i..

Proof. The construction of our Banach algebra A is identical to that of (5.1) with $\underline{H}(n) = 1$ $(n \in Z^+)$, except that the algebra is non-commutative, the relations defining the multiplication being $e(\lambda)e(\mu) =$

$e(\lambda) = e(\mu)e(\lambda)$ $(\lambda \subset \mu)$. Thus a typical element $x \in A$ may be written

$$x = \Sigma\, \xi(\lambda_1,\ldots,\lambda_n)e(\lambda_1)\ldots e(\lambda_n),$$

where the summation is over all sequences $(\lambda_1,\ldots,\lambda_n)$ with no $\lambda_i \subset \lambda_{i+1}$ or $\lambda_{i+1} \subset \lambda_i$, and $||x|| = \Sigma\, |\xi(\lambda_1,\ldots,\lambda_n)| < \infty$. For such x and $\beta_1,\ldots,\beta_n < \omega_2$, we define

$$s(\beta_1,\ldots,\beta_n,x) = \Sigma'\xi(\lambda_{11},\ldots,\lambda_{nk_n})e(\lambda_{11})\ldots e(\lambda_{nk_n}),$$

where the summation Σ' is restricted to the sequences $(\lambda_{11},\lambda_{12},\ldots,\lambda_{1k_1},$ $\lambda_{21},\ldots,\lambda_{nk_n})$ with $k_i \geq 1$ $(1 \leq i \leq n)$ and $\beta_i \in \lambda_{ij}$ $(1 \leq i \leq n,\ 1 \leq j \leq k_i)$.

Now suppose that we have an abelian l.a.i. (not necessarily bounded). As in the proof of (5.1), we construct a sequence of ordinals α_n and a sequence of elements $f_n = u_n + v_n$ of the l.a.i.. (Here we use the fact that the proof of (5.2) did not require the boundedness of $\{f_{\alpha\varepsilon}: \varepsilon > 0,$ $0 \leq \alpha < \omega_2\}$.) As in (19), we prove that

$$s(\alpha_1,\alpha_2,f_1f_2) = s(\alpha_1,\alpha_2,u_1u_2).$$

$$s(\alpha_1,\alpha_2,f_2f_1) = s(\alpha_1,\alpha_2,u_2u_1).$$

However, $s(\alpha_1,\alpha_2,u_2u_1) = 0$, whereas $s(\alpha_1,\alpha_2,u_1u_2) = u_1u_2 \neq 0$, by (20) and (21). Therefore $f_1f_2 \neq f_2f_1$. This completes the proof of the theorem.□

(5.4) Remark. It is possible to replace ω_2 by ω_1 in the examples of (5.1) and (5.3), but at the expense of a more complicated argument: the simple notion of α not occurring in an f_n must be replaced by that of α occurring in f_n to an extent less than ε, as in (4.1).

6. Normed algebras with sequential approximate identities.

In this section we construct an example of a normed algebra which shows that the part of Theorem (3.5) giving an abelian bounded t.a.i. from a sequential bounded t.a.i. may fail in the absence of completeness. We shall call a left (resp. two-sided) approximate identity $\{e_\lambda\}_{\lambda\varepsilon\Lambda}$ semiaccurate if, for all $\mu \varepsilon \Lambda$, there exists $\lambda_o \varepsilon \Lambda$ such that, for all $\lambda \geq \lambda_o$, $e_\lambda e_\mu = e_\mu$ (resp. $e_\lambda e_\mu = e_\mu = e_\mu e_\lambda$).

The methods used in Sections 4 and 5 always produce an algebra with semiaccurate approximate identity. Since every sequential semiaccurate t.a.i. in a normed algebra has an abelian subsequence, our counterexample must not contain a semiaccurate t.a.i.. Because of this we need a new type of construction.

(6.1) Theorem. There exists a normed algebra with a sequential bounded t.a.i. of bound 1, but with no abelian l.a.i..

Proof. Let $(A,||\cdot||_1)$ be the free normed algebra on generators $e(n)$ $(n \varepsilon Z^+)$ with norms $||e(i_1)\ldots e(i_n)||_1 = 1$. Thus, a typical $x \varepsilon A$ may be written

$$x = \Sigma\, \lambda(i_1,\ldots,i_n)e(i_1)\ldots e(i_n), \tag{22}$$

the summation being over all sequences $(i_1,\ldots,i_n)$ but only finitely many of the scalars $\lambda(i_1,\ldots,i_n)$ being nonzero. For such x,

$$||x||_1 = \Sigma\, |\lambda(i_1,\ldots,i_n)|.$$

We say that a positive integer i occurs in x if i is a member of some sequence

$(i_1,\dots,i_n)$ with $\lambda(i_1,\dots,i_n) \neq 0$. Thus the set $O(x)$ of all positive integers occurring in x is finite.

Let $(A_1, ||.||_1)$ denote $(A, ||.||_1)$ with an identity 1 adjoined in the usual way. Let $(X, ||.||)$ be the normed space of all functions $\xi : Z^+ \to A_1$ with $||\xi|| = \sum_{n=1}^{\infty} ||\xi(n)||_1 < \infty$. We define a homomorphism ψ from A into the algebra $L(X)$ of operators on X by

$$[\psi(e(m))\xi](n) = \begin{cases} \xi(n), & \text{if } n < m \\ n^{-1}e(m)\xi(n), & \text{if } n \geq m. \end{cases}$$

We show that ψ is injective. Let x be a nonzero element in A with expression as in (22). Let $N = \max O(x)$, and define $\xi_N \in X$ by $\xi_N(N) = 1$, $\xi_N(m) = 0$ $(m \neq N)$. Then

$$[\psi(x)\xi_N](N) = \Sigma\, n^{-N}\lambda(i_1,\dots,i_n)e(i_1)\dots e(i_n) \neq 0, \qquad (*)$$

so $\psi(x) \neq 0$. We define our desired norm on A by $||a|| = ||\psi(a)||$ for $a \in A$. To show that $\{e(n)\}$ is a l.a.i. in $(A, ||.||)$, we observe that, for $i > j$ and $\xi \in X$,

$$[\psi(e(i)e(j) - e(j))\xi](n) = \begin{cases} 0, & \text{if } n < i, \\ [n^{-2}e(i)e(j) - n^{-1}e(j)]\xi(n), & \text{if } n \geq i, \end{cases}$$

$$||[\psi(e(i)e(j) - e(j))\xi](n)||_1 = \begin{cases} 0, & \text{if } n < i, \\ \leq 2n^{-1}||\xi(n)||_1, & \text{if } n \geq i, \end{cases}$$

$$||\psi(e(i)e(j) - e(j))\xi|| \leq 2i^{-1}||\xi||,$$

$$||e(i)e(j) - e(j)|| < 2i^{-1}.$$

Similarly, we have

$$||e(j)e(i) - e(j)|| \leq 2i^{-1}.$$

It follows that $\{e(n)\}$ is a two-sided approximate identity for A, and clearly $||e(n)|| = 1$ for all $n \in Z^+$.

Suppose that $F \subset A$ is a left approximate identity. Choose $f_1 \in F$, $f_1 \neq 0$, arbitrarily. Let $N = \max O(f_1)$, and let $f_2 \in F$ be such that $||f_2e(N) - e(N)|| < 1$. Define ξ_N as before. Then

$$||\psi(f_2e(N) - e(N))\xi_N|| < 1.$$

Hence $O(f_2) \not\subset [1,N]$, for, otherwise, the coefficient of the monomial $e(N)$ in $[\psi(f_2e(N) - e(N))\xi_N](N)$ would be -1 (see (*) above.)

Now, it is easy to deduce that $f_1f_2 \neq f_2f_1$: for instance, if we consider monomials of f_1f_2 and of f_2f_1 of the form

$$e(i_1)e(i_2)\ldots e(i_n)$$

with $i_1,i_2,\ldots,i_{J-1} \leq N$ and $i_J > N$, then the largest J attainable amongst the monomials of f_1f_2 will be strictly larger than that attainable in f_2f_1. This completes the proof.□

7. <u>Quotients</u>.

Let A be a normed algebra and I a closed two-sided ideal in A. Clearly, if A has a (bounded) l.a.i. $\{e_\lambda\}$, then $\{e_\lambda + I\}$ is a (bounded) l.a.i. of the quotient algebra A/I provided with the quotient norm.

§7. QUOTIENTS

(7.1) Proposition. Let A be a normed algebra and I a closed two-sided ideal in A. If I has a bounded l.a.i. and if A/I has a (bounded) l.a.i., then A has a (bounded) l.a.i.

Proof. Let $\{e_\gamma\}_{\gamma\varepsilon\Gamma}$ be a l.a.i. of I bounded by K and $\{f_\delta + I\}_{\delta\varepsilon\Delta}$ a (bounded) l.a.i. of A/I. (If $\{f_\delta + I\}_{\delta\varepsilon\Delta}$ is bounded, then, without loss of generality, $\{f_\delta\}_{\delta\varepsilon\Delta}$ can be assumed to be bounded in A; for, if $||f_\delta + I|| < H$ for some constant H, then there is a $y_\delta \varepsilon I$ such that $||f_\delta + y_\delta|| < H$.)

Denote by Λ the directed set of all pairs $\lambda = (F,n)$, where F is a finite subset of A and $n = 1,2,\ldots$. Then for each $\lambda = (F,n)$ in Λ with $F = \{x_1,\ldots,x_m\}$, there is an f_{δ_λ} such that

$$||f_{\delta_\lambda} x_i - x_i + I|| < \frac{1}{2(1+K)n} \quad (i = 1,\ldots,m).$$

Hence there are $y_i \varepsilon I$ $(i = 1,\ldots,m)$ such that

$$||f_{\delta_\lambda} x_i - x_i + y_i|| < \frac{1}{2(1+K)n} \quad (i = 1,\ldots,m).$$

Now choose $\gamma_\lambda \varepsilon \Gamma$ so that

$$||e_{\gamma_\lambda} y_i - y_i|| < \frac{1}{2n} \quad (i = 1,\ldots,m).$$

Then

$$\begin{aligned}||(e_{\gamma_\lambda} + f_{\delta_\lambda} - e_{\gamma_\lambda} f_{\delta_\lambda})x_i - x_i|| &\leq ||f_{\delta_\lambda} x_i - x_i + y_i|| + ||e_{\gamma_\lambda} y_i - y_i|| \\ &\quad + ||e_{\gamma_\lambda}||\cdot||f_{\delta_\lambda} x_i - x_i + y_i|| \\ &< (1 + K)\frac{1}{2(1+K)n} + \frac{1}{2n} = \frac{1}{n}\end{aligned}$$

for all $i = 1,\dots,m$. Thus

$$\{e_{\gamma_\lambda} + f_{\delta_\lambda} - e_{\gamma_\lambda} f_{\delta_\lambda}\}_{\lambda\varepsilon\Lambda}$$

is a (bounded) l.a.i. of A. □

The preceding proof shows that if I has a l.a.i. of norm K and A/I a l.a.i. of norm H, then A has a l.a.i. of norm at most $H + K + HK$. The next example demonstrates that this is the best possible result.

(7.2) Example. Let A be the commutative algebra generated by two elements e, f with the relations $e^2 = e$, $f^2 = f$. Clearly, A is 3-dimensional with basis $\{e, f, ef\}$. For $H, K \geq 1$ consider the algebra norm on A defined by

$$||\alpha_1 e + \alpha_2 f + \alpha_3 ef|| = K|\alpha_1| + H|\alpha_2| + HK|\alpha_3|.$$

Then A is a Banach algebra with respect to this norm. The set $I = Ce + Cef$ is a closed two-sided ideal in A with identity e of norm K. Furthermore, $f + I$ is an identity of the quotient algebra A/I of norm H. The identity $e + f - ef$ of A has norm $H + K + HK$.

8. Tensor Products.

In this section we will study the existence of approximate identities in the tensor product $A\otimes B$ of two normed algebras versus their existence in A and B.

§8. TENSOR PRODUCTS

First let us gather some facts from the general theory of tensor products.

Let E, F be normed linear spaces and $E \otimes F$ their <u>algebraic tensor product</u>. If $||\cdot||_\alpha$ is a norm on $E \otimes F$, then the normed linear space $(E \otimes F, ||\cdot||_\alpha)$ will be denoted by $E \otimes^\alpha F$. The completion of $E \otimes^\alpha F$ is denoted by $E \otimes_\alpha F$.

Let $t \in E \otimes_\alpha F$; then for every $\varepsilon > 0$ there is a representation

$$t = \sum_{i=1}^{\infty} t_i \quad \text{with} \quad t_i \in E \otimes F$$

and

$$\sum_{i=1}^{\infty} ||t_i||_\alpha \leq ||t||_\alpha + \varepsilon.$$

Indeed, since there exists a sequence $\{t'_n\}$, where each $t'_n \in E \otimes F$, which converges to t with respect to $||\cdot||_\alpha$, we may assume that $||t - t'_1||_\alpha \leq \frac{\varepsilon}{2}$ and $||t'_{n+1} - t'_n||_\alpha \leq \frac{\varepsilon}{2^{n+1}}$ for all n, from which we get the desired representation $t = t'_1 + \sum_{n=1}^{\infty} (t'_{n+1} - t'_n)$.

A norm $||\cdot||_\alpha$ on $E \otimes F$ is called a <u>cross-norm</u> if

$$||x \otimes y||_\alpha = ||x||\cdot||y|| \quad \text{for all} \quad x \in E,\ y \in F.$$

The <u>greatest cross-norm</u> $||\cdot||_\gamma$ on $E \otimes F$ is given by

$$||t||_\gamma = \inf\{\sum_{i=1}^{n} ||x_i||\cdot||y_i|| : t = \sum_{i=1}^{n} x_i \otimes y_i\}.$$

For the greatest cross-norm some formulas become considerably simpler.

Let $t \varepsilon E \otimes_\gamma F$; then there is a representation

$$t = \sum_{i=1}^{\infty} x_i \otimes y_i \quad \text{with} \quad x_i \varepsilon E,\ y_i \varepsilon F,$$

and

$$\sum_{i=1}^{\infty} ||x_i|| \cdot ||y_i|| < \infty.$$

Also

$$||t||_\gamma = \inf \sum_{i=1}^{\infty} ||x_i|| \cdot ||y_i||,$$

where the infimum is taken over all representations of t.

If ϕ, ψ are linear functionals on E and F respectively, then the linear functional $\phi \otimes \psi$ on $E \otimes F$ is defined by

$$(\phi \otimes \psi)(x \otimes y) = \phi(x)\psi(y).$$

The <u>least</u> <u>cross-norm</u> (whose associate [241, p. 26] is also a cross-norm) $||\cdot||_\lambda$ on the tensor product of E and F is defined as follows:

$$||t||_\lambda = \sup |(\phi \otimes \psi)(t)|,$$

where the supremum is taken over all continuous linear functionals ϕ, ψ on E and F respectively, with

$$||\phi|| = 1 = ||\psi||.$$

Clearly $||\cdot||_\lambda \leq ||\cdot||_\gamma$.

A norm $||\cdot||_\alpha$ on $E \otimes F$ will be called <u>admissible</u>, if there exist positive constants M_λ, M_γ, such that

$$M_\lambda ||\cdot||_\lambda \leq ||\cdot||_\alpha \leq M_\gamma ||\cdot||_\gamma .$$

Clearly $M_\lambda \leq M_\gamma$.

If A, B are normed algebras, then $A \otimes B$ is an algebra with respect to the multiplication defined by

$$(x_1 \otimes y_1)(x_2 \otimes y_2) = x_1 x_2 \otimes y_1 y_2 .$$

The greatest cross-norm $||\cdot||_\gamma$ is always an algebra norm on $A \otimes B$; but for most algebras the least cross-norm $||.||_\lambda$ is not submultiplicative.

Now let us investigate the existence of approximate identities in tensor products.

(8.1) <u>Proposition</u>. Let A, B be normed algebras and $||\cdot||_\alpha$ an admissible algebra norm on $A \otimes B$. If A, B have (bounded) l.a.i.s $\{e_\mu\}_{\mu \in M}$ and $\{f_\nu\}_{\nu \in N}$ respectively, then $\{e_\mu \otimes f_\nu\}_{(\mu,\nu) \in M \times N}$ is a (bounded) l.a.i. of $A \otimes^\alpha B$.

<u>Proof</u>. For any $t = \sum_{i=1}^n x_i \otimes y_i$ in $A \otimes B$,

$$t - (e_\mu \otimes f_\nu)t = \sum_{i=1}^n x_i \otimes y_i - \sum_{i=1}^n e_\mu x_i \otimes f_\nu y_i$$

$$= \sum_{i=1}^n (x_i - e_\mu x_i) \otimes y_i + \sum_{i=1}^n x_i \otimes (y_i - f_\nu y_i)$$

$$- \sum_{i=1}^n (x_i - e_\mu x_i) \otimes (y_i - f_\nu y_i),$$

and so

$$||t - (e_\mu \otimes f_\nu)t||_\alpha \le \sum_{i=1}^{n} ||(x_i - e_\mu x_i)\otimes y_i||_\alpha + \sum_{i=1}^{n} ||x_i \otimes (y_i - f_\nu y_i)||_\alpha$$

$$+ \sum_{i=1}^{n} ||(x_i - e_\mu x_i) \otimes (y_i - f_\nu y_i)||_\alpha$$

$$\le M_\gamma \sum_{i=1}^{n} ||x_i - e_\mu x_i||\cdot||y_i|| + M_\gamma \sum_{i=1}^{n} ||x_i||\cdot||y_i - f_\nu y_i||$$

$$+ M_\gamma \sum_{i=1}^{n} ||x_i - e_\mu x_i||\cdot||y_i - f_\nu y_i||$$

Hence

$$\lim_{(\mu,\nu)\in M\times N} ||t - (e_\mu \otimes f_\nu)t||_\alpha = 0.$$

If $\{e_\mu\}_{\mu\in M}$ and $\{f_\nu\}_{\nu\in N}$ are bounded by H and K respectively, then $||e_\mu \otimes f_\nu||_\alpha \le M_\gamma||e_\mu||\cdot||f_\nu|| \le M_\gamma HK$; i.e., $\{e_\mu \otimes f_\nu\}_{(\mu,\nu)\in M\times N}$ is bounded by $M_\gamma HK$ — and, in this case, $\{e_\mu \otimes f_\nu\}_{(\mu,\nu)\in M\times N}$ is also a bounded l.a.i. for the completion $A \otimes_\alpha B$. □

(8.2) <u>Theorem</u>. Let A, B be Banach algebras and $||\cdot||_\alpha$ an admissible algebra norm on $A\otimes B$. If $A \otimes_\alpha B$ has a (bounded) l.a.i., then A and B have (bounded) l.a.i.'s.

<u>Proof</u>. Let $\{t_\nu\}_{\nu\in N}$ be a l.a.i. of $A \otimes_\alpha B$. We show that B has a l.a.i.. A similar argument shows that A also has a l.a.i..

Let $I : B \to B$ denote the identity operator on B. For $\psi \in A^*$

the relation $(\psi \otimes I)(a \otimes b) = \psi(a)b$ defines a linear map $A \otimes B \to B$ which is $||.||_\lambda$-continuous, and so it extends to $A \otimes_\lambda B$ because, for $t \in A \otimes B$,

$$\begin{aligned} ||(\psi \otimes I)t|| &= \sup \{|((\psi \otimes I)t, g)| : g \in B^*, ||g|| \leq 1\} \\ &= \sup \{|(\psi \otimes g)(t)| : g \in B^*, ||g|| \leq 1\} \\ &\leq ||\psi|| \cdot ||t||_\lambda . \end{aligned}$$

The identity map on $A \otimes B$ extends to a continuous map $A \otimes_\alpha B \to A \otimes_\lambda B$ and composing this with $\psi \otimes I$ we obtain a continuous map $\psi \otimes_\alpha I$ from $A \otimes_\alpha B$ into B. For $a \in A$ fixed, define $T_a : A \to A$ by $T_a x = xa$.

If $\phi \in A^*$, then

$$[(T_a^* \phi \otimes_\alpha I)(c \otimes d)]b = (\phi \otimes_\alpha I)((c \otimes d)(a \otimes b)).$$

So, by linearity and continuity we have

$$(T_a^* \phi \otimes_\alpha I)(t)b = (\phi \otimes_\alpha I)(t(a \otimes b)), \quad t \in A \otimes_\alpha B.$$

Hence, choosing a nonzero $a \in A$ and $\phi \in A^*$ with $||\phi|| = ||a|| = \phi(a) = 1$, we obtain

$$\begin{aligned} (T_a^* \phi \otimes_\alpha I)(t_\nu)b &= (\phi \otimes_\alpha I)(t_\nu(a \otimes b)) \\ &\to (\phi \otimes_\alpha I)(a \otimes b) = b \end{aligned}$$

for all $b \in B$. Therefore $\{f_\nu\}_{\nu \in N}$, where $f_\nu = (T_a^* \phi \otimes_\alpha I)(t_\nu)$, is a l.a.i. for B which is bounded if $\{t_\nu\}_{\nu \in N}$ is.□

(8.3) <u>Corollary</u>. Let A, B be normed algebras and $||.||_\alpha$ an admissible algebra norm on $A \otimes B$. If $A \otimes^{\alpha} B$ has a (bounded) l.a.i., then A and B have (bounded) l.a.i.s.

9. Approximate units.

Motivated by the criteria in Propositions (1.2) and (1.3) for the existence of an approximate identity in a normed algebra, we introduce the more general concept of approximate units.

(9.1) Definitions. A normed algebra A has left (resp. right, two-sided) approximate units, abbreviated l.a.u. (resp. r.a.u., t.a.u.) if, given any $x \in A$ and $\varepsilon > 0$ there is an element $u \in A$ (depending on x and ε) such that

$$||x - ux|| < \varepsilon$$

(resp. $||x - xu|| < \varepsilon$, $||x - ux|| < \varepsilon$ and $||x - xu|| < \varepsilon$).

A normed algebra A has bounded l.a.u. (resp. r.a.u., t.a.u.) if there is a constant K such that for every $x \in A$ and $\varepsilon > 0$ there is an element $u \in A$ (depending on x and ε) with $||u|| \leq K$ such that

$$||x - ux|| < \varepsilon$$

(resp. $||x - xu|| < \varepsilon$, $||x - ux|| < \varepsilon$ and $||x - xu|| < \varepsilon$).

A normed algebra A has pointwise-bounded l.a.u. (resp. r.a.u., t.a.u.) if for every $x \in A$ there is a constant $K = K(x)$ such that for every $\varepsilon > 0$ there is an element $u \in A$ (depending on x and ε) with $||u|| \leq K(x)$ such that

$$||x - ux|| < \varepsilon$$

(resp. $||x - xu|| < \varepsilon$, $||x - ux|| < \varepsilon$ and $||x - xu|| < \varepsilon$).

(9.2) Proposition. A normed algebra A has l.a.u. (resp. r.a.u.) if and only if there exists a positive number $q < 1$ with the following property: for every $x \in A$ there is an element $u \in A$ such that

$$||x - ux|| \leq q||x||$$

(resp. $||x - xu|| \leq q||x||$).

Proof. If A has l.a.u. then for any positive number $q < 1$ and every $x \in A$ there is an element $u \in A$ with

$$||x - ux|| \leq q||x||.$$

Conversely, with the usual convention about the formal role of 1, the assumption takes the following form: for every $x \in A$ there is an element $u \in A$ such that $||(1 - u)x|| \leq q||x||$.

Now let x be any element in A. Then we can choose successively $u_1,\ldots,u_n \in A$ such that

$$||(1 - u_i)\cdots(1 - u_1)x|| \leq q||(1 - u_{i-1})\cdots(1 - u_1)x|| \quad (i = 1,\ldots,n).$$

Define u in A by

$$1 - u = (1 - u_n)\cdots(1 - u_1).$$

Then

$$||x - ux|| \leq q^n||x||.$$

Hence, given any $x \in A$ and $\varepsilon > 0$ there is an element $u \in A$ (depending on x and ε) such that $||x - ux|| < \varepsilon$; i.e., A has l.a.u.□

Obviously, every normed algebra with a bounded l.a.i. has bounded l.a.u. The next theorem states the converse.

(9.3) Theorem. A normed algebra A has l.a.u. bounded by a constant K if and only if A has a l.a.i. bounded by the same constant K.

Proof. Let A be a normed algebra with l.a.u. bounded by a constant K. Let $\{x_1,\ldots,x_n\}$ be any finite set of elements in A. Given $\varepsilon > 0$, we can choose successively $u_1,\ldots,u_n \in A$ such that $||u_i|| \leq K$ and

$$||(1 - u_i)\cdots(1 - u_2)(1 - u_1)x_i|| < \varepsilon \quad (i = 1,\ldots,n).$$

Define v in A by

$$1 - v = (1 - u_n)\cdots(1 - u_2)(1 - u_1).$$

Then

$$||x_i - vx_i|| = ||[(1 - u_n)\cdots(1 - u_{i+1})][(1 - u_i)\cdots(1 - u_1)x_i]||$$

$$\leq (1 + K)^{n-i}||(1 - u_i)\cdots(1 - u_1)x_i|| \leq (1 + K)^n\varepsilon.$$

Finally we choose u in A with $||u|| \leq K$ and $||v - uv|| < \varepsilon$. Then for each $i = 1,2,\ldots,n$ we have:

$$||x_i - ux_i|| \leq ||x_i - vx_i|| + ||(v - uv)x_i|| + ||u(x_i - vx_i)||$$

$$\leq ||x_i - vx_i|| + ||v - uv||\cdot||x_i|| + ||u||\cdot||x_i - vx_i||$$

$$\leq (1 + K)^n\varepsilon + ||x_i||\varepsilon + K(1 + K)^n\varepsilon.$$

Hence, for every finite set $\{x_1,\ldots,x_n\}$ of elements in A and every $\varepsilon > 0$ there is an element u in A such that $||u|| \leq K$ and $||x_i - ux_i|| < \varepsilon$ for all $i = 1,\ldots,n$. By Proposition (1.2) we conclude that the normed algebra A has a l.a.i. bounded by the constant K. □

(9.4) Theorem. A normed algebra A has t.a.u. bounded by a constant K if and only if A has a t.a.i. bounded by the same constant K.

Proof. Let A be a normed algebra with t.a.u. bounded by a constant K. Let $\{x_1,\ldots,x_n\}$ be any finite set of elements in A. Given $\varepsilon > 0$ we can find $u, v \in A$ with $||u|| \leq K$, $||v|| \leq K$ such that

$$||(1 - u)x_i|| < \varepsilon \quad \text{and} \quad ||x_i(1 - v)|| < \varepsilon \quad \text{for all } i = 1,\ldots,n.$$

Define w in A by $1 - w = (1 - v)(1 - u)$. Then

$$||(1 - w)x_i|| = ||(1 - v)(1 - u)x_i|| \leq ||1 - v||\cdot||(1 - u)x_i|| < (1 + K)\varepsilon$$

and

$$||x_i(1 - w)|| = ||x_i(1 - v)(1 - u)|| \leq ||x_i(1 - v)||\cdot||1 - u|| < (1 + K)\varepsilon$$

for all $i = 1,\ldots,n$. Finally we choose e in A with $||e|| \leq K$, $||w - ew|| < \varepsilon$, and $||w - we|| < \varepsilon$. Then for each $i = 1,\ldots,n$ we have:

$$\begin{aligned} ||x_i - ex_i|| &\leq ||x_i - wx_i|| + ||(w - ew)x_i|| + ||e(x_i - wx_i)|| \\ &\leq ||x_i - wx_i|| + ||w - ew||\cdot||x_i|| + ||e||\cdot||x_i - wx_i|| \\ &\leq (1 + K)\varepsilon + ||x_i||\varepsilon + K(1 + K)\varepsilon \end{aligned}$$

and

$$||x_i - x_ie|| \leq ||x_i - x_iw|| + ||x_i(w - we)|| + ||(x_i - x_iw)e||$$

$$\leq ||x_i - x_iw|| + ||x_i||\cdot||w - we|| + ||x_i - x_iw||\cdot||e||$$

$$\leq (1 + K)\varepsilon + ||x_i||\varepsilon + K(1 + K)\varepsilon.$$

Hence, for every finite set $\{x_1,\dots,x_n\}$ of elements in A and every $\varepsilon > 0$ there is an element e in A such that $||e|| \leq K$, $||x_i - ex_i|| < \varepsilon$, and $||x_i - x_ie|| < \varepsilon$ for all $i = 1,\dots,n$. By Proposition (1.3) we conclude that the normed algebra A has a t.a.i. bounded by the constant K. □

(9.5) <u>Proposition</u>. A commutative normed algebra A with pointwise-bounded approximate units has an approximate identity (possibly unbounded).

<u>Proof</u>. Let A be a commutative normed algebra with pointwise-bounded approximate units. Let $\{x_1,\dots,x_n\}$ be any finite set of elements in A. Set $K = K(x_1,\dots,x_n) = \max\{K(x_1),\dots,K(x_n)\}$. Given $\varepsilon > 0$, there are elements $u_1,\dots,u_n$ in A such that $||u_i|| \leq K$ and $||x_i - u_ix_i|| < \varepsilon$ for all $i = 1,\dots,n$. Define u in A by

$$1 - u = (1 - u_n)\cdots(1 - u_1).$$

Then

$$||x_i - ux_i|| = ||(1 - u_n)\cdots(1 - u_1)x_i||$$

$$\leq (1 + K)^{n-1} ||(1 - u_i)x_i|| \leq (1 + K)^{n-1}\varepsilon.$$

Hence, for every finite set $\{x_1,\ldots,x_n\}$ of elements in A and every $\varepsilon > 0$ there is an element u in A such that $||x_i - ux_i|| < \varepsilon$ for all $i = 1,\ldots,n$. By Proposition (1.2) the assertion follows.□

The next example shows that we cannot in general assert the existence of a bounded approximate identity.

(9.6) Example. Consider the commutative normed algebra

$$A = \{(\alpha_1,\alpha_2,\ldots): \alpha_i \text{ complex and } \alpha_i = 0 \text{ for almost all } i\}$$

with coordinatewise algebraic operations and norm defined by

$$||(\alpha_1,\alpha_2,\ldots)|| = \max |i\alpha_i|.$$

Then A has pointwise-bounded approximate units

$$u_i = (1,\ldots,1,0,0,\ldots).$$

Obviously, A has no bounded approximate identity.

(9.7) Theorem. A commutative Banach algebra A has pointwise-bounded approximate units if and only if A has a bounded approximate identity.

Proof. Let A be a commutative Banach algebra with pointwise-bounded approximate units. Define for each $n = 1,2,\ldots$

$A_n = \{x \in A: \lim u_i x = x$ for some sequence $\{u_i\}$ in A with $||u_i|| \leq n\}$.

Then A_n is a closed subset of A. Indeed, if $\{x_j\}$ is a sequence in A_n with $\lim x_j = x$, then there exist sequences $\{u_{ij}\}_i$ in A such that $||u_{ij}|| \leq n$ and $\lim_i u_{ij}x_j = x_j$. Then

$$||x - u_{ij}x|| \leq ||x - x_j|| + ||x_j - u_{ij}x_j|| + ||u_{ij}x_j - u_{ij}x||$$

$$\leq ||x - x_j|| + ||x_j - u_{ij}x_j|| + ||u_{ij}||\cdot||x_j - x||$$

$$\leq (1 + n)||x - x_j|| + ||x_j - u_{ij}x_j||;$$

choosing first j and then i large enough, it follows that $||x - u_{ij}x||$ can be made arbitrarily small. Hence $x \in A_n$; i.e., A_n is closed.

Since A is the union of the sets A_n, $n = 1,2,\ldots$, and A is a Banach space, it follows from the Baire category theorem, that some A_n has nonempty interior. Thus $B(x_o,\delta) = \{x \in A: ||x - x_o|| < \delta\}$ is a subset of A_{n_o} for some $x_o \in A$, $\delta > 0$ and n_o. We will show that $B(0,\delta) = \{x \in A: ||x|| < \delta\}$ is a subset of $A_{(2+n_o)n_o}$. Let $x \in B(0,\delta)$; then $x = (x + x_o) - x_o$ with $x + x_o$ and x_o in $B(x_o,\delta)$. Hence there exist sequences $\{u_i\}$ and $\{v_i\}$ in A such that $||u_i|| \leq n_o$, $||v_i|| \leq n_o$, $\lim u_i(x + x_o) = x + x_o$ and $\lim v_i x_o = x_o$. Set $w_i = u_i + v_i - u_i v_i$; then $\{w_i\}$ is a sequence in A with $||w_i|| \leq (2 + n_o)n_o$ and $\lim w_i x = x$; i.e., x is in $A_{(2+n_o)n_o}$.

Since $\alpha\cdot A_{(2+n_o)n_o}$ is a subset of $A_{(2+n_o)n_o}$ for any scalar α, it follows that $A_{(2+n_o)n_o} = A$; i.e., A has bounded approximate units and so, by Theorem (9.3), A has a bounded approximate identity. □

10. Topological zero divisors.

Here we study approximate units and approximate identities in normed algebras which do not consist entirely of topological zero divisors.

Let A be a normed algebra. An element z in A is called a <u>right topological zero divisor</u> in A if there is a sequence $\{x_n\}$ of elements in A with $||x_n|| = 1$ for all n such that $\lim x_n z = 0$. The set of all right topological zero divisors in A will be denoted by $Z^r(A)$. Then $Z^r(A)$ is closed in A so that the complementary set $H^r(A) = A \setminus Z^r(A)$ is open.

Let $\tilde{A}$ be the completion of A. Then $H^r(A) \subset H^r(\tilde{A})$. This may be seen as follows: Assume a is in $H^r(A) \cap Z^r(\tilde{A})$. Then there exists a sequence $\{\tilde{x}_n\}$ of elements in $\tilde{A}$ such that $||\tilde{x}_n|| = 1$ and $\lim \tilde{x}_n a = 0$. Since A is a dense subspace of $\tilde{A}$ there is a sequence $\{x_n\}$ of elements in A such that $||x_n|| = 1$ and $\lim (x_n - \tilde{x}_n) = 0$. Then $\lim x_n a = \lim (x_n - \tilde{x}_n)a + \lim \tilde{x}_n a = 0$. Contradiction! Hence $H^r(A) \cap Z^r(\tilde{A}) = \emptyset$ and so $H^r(A) \subset H^r(\tilde{A})$.

(10.1) <u>Definition</u>. Let A be a normed algebra. A net $\{u_\lambda\}$ in A is called a <u>uniform left</u> (resp. <u>right</u>, <u>two-sided</u>) <u>approximate identity in</u> A if $||x - u_\lambda x||$ (resp. $||x - xu_\lambda||$, $||x - u_\lambda x||$ and $||x - xu_\lambda||$) converges uniformly to zero for all x in A with $||x|| = 1$.

Obviously, if u is a right identity in A then u is a uniform right approximate identity in A. Furthermore u is in $H^r(A)$ and so $A \neq Z^r(A)$.

(10.2) Lemma. Let A be a normed algebra with a uniform right approximate identity. Then $A \neq Z^r(A)$.

Proof. Let $\{u_\lambda\}$ be a uniform right approximate identity in A. Fix a in $\{u_\lambda\}$ such that

$$||x - xa|| < \frac{1}{2} \text{ for all } x \in A \text{ with } ||x|| = 1.$$

Then a is not a right topological zero divisor in A.□

(10.3) Lemma. Let A be a normed algebra with $A \neq Z^r(A)$. Let a be an element in $H^r(A)$ and assume that there is a net $\{u_\lambda\}$ of elements in A such that $\lim u_\lambda a = a$. Then $\{u_\lambda\}$ is a right approximate identity in A.

Proof. Let x be any element in A. Then $\lim xu_\lambda a = xa$. Since a is not a right topological zero divisor in A it follows that $\lim xu_\lambda = x$.□

(10.4) Proposition. Let A be a normed algebra with $A \neq Z^r(A)$. If A has left approximate units then the completion of A has a right identity, and so A has a bounded uniform right approximate identity.

Proof. Fix a in $H^r(A)$ and choose u_n in A such that

$$||a - u_n a|| < \frac{1}{n}, \quad n = 1,2,\ldots .$$

Then

$$||(u_n - u_m)a|| < \frac{1}{n} + \frac{1}{m}, \quad n, m = 1,2,\ldots .$$

Since a is not a right topological zero divisor in A it follows that $\{u_n\}$ is a Cauchy sequence in A. Hence $u = \lim u_n$ is a right identity in the completion of A and $\{u_n\}$ is a bounded uniform right approximate identity in A. □

(10.5) Proposition. Let A be a normed algebra with a right identity. If A has left approximate units then A has a two-sided identity.

Proof. Let e be a right identity in A. Then for any x in A there exists a sequence $\{u_n\}$ of elements in A such that

$$\lim u_n(x - ex) = x - ex.$$

Since

$$u_n(x - ex) = u_n x - u_n ex = u_n x - u_n x = 0$$

it follows that

$$x - ex = 0.$$

Thus e is also a left identity in A. □

(10.6) Theorem. Let A be a Banach algebra with $A \neq Z^r(A)$. If A has left approximate units then A has a two-sided identity.

Proof. By Proposition (10.4) A has a right identity, and so by

Proposition (10.5) it also has a two-sided identity.□

(10.7) Proposition. Let A be a normed algebra with $A \neq Z^r(A)$. If A has a left approximate identity then the completion of A has a two-sided identity, and so A has a bounded uniform two-sided approximate identity.

Proof. Fix a in $H^r(A)$. Then, as in the proof of (10.4), A has a bounded uniform right approximate identity $\{u_n\}$ in A, and $u = \lim u_n$ is a right identity in the completion of A.

Now let $\{e_\lambda\}$ be any left approximate identity in A. Then

$$\lim e_\lambda a = a = \lim u_n a = ua.$$

Since a is not a right topological zero divisor in the completion of A it follows that

$$\lim e_\lambda = u.$$

Thus $ux = \lim e_\lambda x = x$ for all x in A. Hence $u = \lim u_n$ is a left identity in the completion of A and $\{u_n\}$ is a bounded uniform left approximate identity in A.□

11. Topologically nilpotent elements.

Let A be a normed algebra. For $x \in A$ define

$$\nu(x) = \lim ||x^n||^{1/n}.$$

If $\nu(x) = 0$ then x is called topologically nilpotent. The set of all

§11. TOPOLOGICALLY NILPOTENT ELEMENTS

topogically nilpotent elements in A is denoted by $N(A)$. The following lemma is well known (see Rickart [222, pages 24-25] for a proof).

(11.1) Lemma. Let A be a normed algebra all of whose elements are topologically nilpotent. Then every element in A is a two-sided topological zero divisor.

(11.2) Proposition. Let A be a normed algebra with $A = N(A)$. Then A cannot have a uniform right approximate identity.

Proof. Assume A has a uniform right approximate identity. Then $A \neq Z^r(A)$ by Lemma (10.2). But by Lemma (11.1), $A = N(A)$ implies $A = Z^r(A)$. Contradiction! □

The next example demonstrates that a Banach algebra A with $A = N(A)$ may still have a bounded approximate identity.

(11.3) Example. Consider the space $A = L(0,1)$ of all complex valued Lebesgue integrable functions on the interval $[0,1]$. Under pointwise addition and multiplication by scalars and the norm

$$||f|| = \int_0^1 |f(x)|\, dx,$$

A is a Banach space. It becomes a commutative Banach algebra under the convolution multiplication,

$$(f*g)(x) = \int_0^x f(y)g(x-y)\, dy \quad \text{for } f, g \in A.$$

It is easy to check that

$$||f*\dots*f|| \le \frac{1}{n!} \sup_{x\varepsilon[0,1]} |f(x)|^n$$

for all continuous $f \varepsilon A$. Since the continuous functions are dense in A it follows that $A = N(A)$.

On the other hand the sequence $\{u_n\}_{n=1}^{\infty}$ of functions in A defined by

$$u_n(x) = \begin{cases} n, & \text{if } 0 \le x < \frac{1}{n}, \\ 0, & \text{if } \frac{1}{n} \le x \le 1, \end{cases}$$

is an approximate identity in A bounded by one. Indeed, for every continuous function $f \varepsilon A$ and any $\varepsilon > 0$,

$$|f(x) - f(x - y)| < \varepsilon \quad \text{for} \quad x \varepsilon [\tfrac{1}{n},1],\ y \varepsilon [0,\tfrac{1}{n}],$$

for all sufficiently large n, and so:

$$||f - u_n*f|| = \int_0^1 |f(x) - \int_0^x u_n(y)f(x - y)\ dy|dx$$

$$= \int_0^{\frac{1}{n}} |f(x) - n\int_0^x f(x - y)\ dy|dx + \int_{\frac{1}{n}}^1 |f(x) - n\int_0^{\frac{1}{n}} f(x - y)\ dy|dx$$

$$\le \int_0^{\frac{1}{n}} |f(x)|\ dx + n\int_0^{\frac{1}{n}}\int_0^x |f(x - y)|\ dy\ dx$$

$$+ n\int_{\frac{1}{n}}^1\int_0^{\frac{1}{n}} |f(x) - f(x - y)|\ dy\ dx$$

$$\leq \varepsilon + \frac{2}{n} \cdot \sup_{x \in [0,1]} |f(x)|.$$

Thus $\lim_n u_n * f = f$ for all continuous $f \in A$. Since the continuous functions are dense in A and $\{u_n\}_{n=1}^{\infty}$ is bounded, it follows that $\{u_n\}_{n=1}^{\infty}$ is an approximate identity in A. □

(11.4) Definition. A normed algebra A is called topologically nilpotent if

$$\lim_n \sup_{||x|| \leq 1} ||x^n||^{1/n} = 0.$$

Obviously, if A is topologically nilpotent then the completion of A is also topologically nilpotent.

If A is any normed linear space then A with trivial multiplication $xy = 0$ for all x and y in A is a commutative, topologically nilpotent, normed algebra which has no approximate units at all.

There are also examples of topologically nilpotent Banach algebras with unbounded approximate identities.

(11.5) Example. Consider the space $A = C(0,1)$ of all complex valued continuous functions on the interval $[0,1]$. Under pointwise addition and multiplication by scalars and the sup-norm

$$||f||_{\infty} = \sup_{x \in [0,1]} |f(x)|,$$

A is a Banach space. It becomes a Banach algebra under the convolution multiplication,

$$(f*g)(x) = \int_o^x f(y)g(x-y)\,dy \quad \text{for} \quad f,\ g \in A,$$

which is commutative.

It is easy to check that

$$||f*\ldots*f||_\infty \leq \frac{1}{(n-1)!}\,||f||_\infty^n.$$

for all $f \in A$. Thus A is topologically nilpotent.

Now consider the subalgebra

$$A_o = \{f \in A\colon f(0) = 0\},$$

which is also topologically nilpotent.

On the other hand the sequence $\{u_n\}_{n=1}^\infty$ of functions in A_o defined by

$$u_n(x) = \begin{cases} \frac{n\pi}{2}\cdot \sin n\pi x, & \text{if} \quad 0 \leq x \leq \frac{1}{n}, \\ 0, & \text{if} \quad \frac{1}{n} < x \leq 1, \end{cases}$$

is an approximate identity in A_o. Indeed, for every $f \in A_o$ and any $\varepsilon > 0$,

$$|f(x)| < \varepsilon \quad \text{for} \quad x \in [0,\tfrac{1}{n}]$$

and

$$|f(x) - f(x-y)| < \varepsilon \quad \text{for} \quad x \in [\tfrac{1}{n},1],\ y \in [0,\tfrac{1}{n}],$$

for all sufficiently large n, and so

$$||f - u_n*f||_\infty = \sup_{x\varepsilon[0,1]} |f(x) - \int_o^x u_n(y)f(x-y)\,dy|$$

$$\le \sup_{x\varepsilon[0,\frac{1}{n}]} \left|f(x) - \int_0^x \frac{n\pi}{2}\cdot\sin(n\pi y)f(x-y)\,dy\right|$$

$$+ \sup_{x\varepsilon[\frac{1}{n},1]} \left|f(x) - \int_0^{\frac{1}{n}} \frac{n\pi}{2}\cdot\sin(n\pi y)f(x-y)\,dy\right|$$

$$\le \sup_{x\varepsilon[0,\frac{1}{n}]} |f(x)| + \sup_{x\varepsilon[0,\frac{1}{n}]} \int_0^x \frac{n\pi}{2}\cdot\sin(n\pi y)|f(x-y)|\,dy$$

$$+ \sup_{x\varepsilon[\frac{1}{n},1]} \int_0^{\frac{1}{n}} \frac{n\pi}{2}\cdot\sin(n\pi y)|f(x) - f(x-y)|\,dy$$

$$\le 3\varepsilon.$$

Thus $\lim_n u_n * f = f$ for all continuous $f \varepsilon A$. Hence $\{u_n\}_{n=1}^{\infty}$ is an approximate identity in A which is clearly unbounded in the sup-norm.□

However, as we shall see next, a topologically nilpotent algebra cannot have a bounded approximate identity.

(11.6) Proposition. Let A be a topologically nilpotent normed algebra. Then A cannot have pointwise-bounded right approximate units.

Proof. Assume that there exists a nonzero element x in A and a sequence $\{u_i\}$ of elements in A bounded by some constant $K \geq 1$ such that $\lim xu_i = x$. Then

$$||x - xu_i^n|| \leq ||x - xu_i|| + ||xu_i - xu_i^2|| + \cdots + ||xu_i^{n-1} - xu_i^n||$$

$$\leq ||x - xu_i||(1 + ||u_i|| + \cdots + ||u_i^{n-1}||)$$

$$\leq ||x - xu_i||(1 + K + \cdots + K^{n-1}),$$

so that

$$\lim_i xu_i^n = x \quad \text{for each} \quad n = 1,2,\ldots .$$

Since

$$||x|| = \lim_i ||xu_i^n|| \leq \sup_i ||x||\cdot||u_i^n||,$$

it follows that

$$1 \leq \sup_i ||u_i^n||,$$

and so

$$1 \leq \sup_i ||u_i^n||^{1/n} \quad \text{for each} \quad n = 1,2,\ldots .$$

Contradiction! Thus A cannot have pointwise-bounded right approximate units. □

12. C^*-algebras.

Here we shall study approximate identities in C^*-algebras. The most important result is that every C^*-algebra has a two-sided approximate identity bounded by one.

§12. C^*-ALGEBRAS

Let A be a normed *-algebra with continuous involution. If $\{e_\lambda\}_{\lambda\varepsilon\Lambda}$ is a (bounded) left approximate identity in A, then $\{e_\lambda^*\}_{\lambda\varepsilon\Lambda}$ is a (bounded) right approximate identity. If $\{e_\lambda\}_{\lambda\varepsilon\Lambda}$ is a (bounded) two-sided approximate identity in A, then $\{\frac{1}{2}(e_\lambda^* + e_\lambda)\}_{\lambda\varepsilon\Lambda}$ is also a (bounded) two-sided approximate identity in A.

If $\{e_\lambda\}_{\lambda\varepsilon\Lambda}$ is a bounded two-sided approximate identity in A, then $\{e_\lambda^* e_\lambda\}_{\lambda\varepsilon\Lambda}$ is also a bounded two-sided approximate identity in A.

A C^*-_algebra_ is a Banach *-algebra satisfying the C^*-condition

$$||x^*x|| = ||x||^2.$$

The Banach space $C_o(S)$ of continuous complex valued functions which vanish at infinity on a _locally compact Hausdorff space_ S is easily seen to be a commutative C^*-algebra under pointwise multiplication $(fg)(s) = f(s)g(s)$, involution $f^*(s) = \overline{f(s)}$, and sup-norm. By the Gelfand-Naimark theorem every commutative C^*-algebra is isometrically *-isomorphic to some $C_o(S)$ (see [70, 1.4.1]).

(12.1) _Proposition_. The commutative C^*-algebra $C_o(S)$ has an identity element if and only if S is compact.

Proof. If S is compact then the function $e(s) = 1$ $(s \varepsilon S)$ is an identity element in $C_o(S)$.

Conversely, assume $C_o(S)$ has an identity e. Fix any $s_o \varepsilon S$. Then there exists a function $f \varepsilon C_o(S)$ with $f(s_o) = 1$. Since e is an identity in $C_o(S)$ it follows that

$$e(s_o) = e(s_o)f(s_o) = f(s_o) = 1.$$

Since $s_o \in S$ was arbitrary, $e(s) = 1$ for all $s \in S$. Thus S is compact.

(12.2) <u>Proposition</u>. The C^*-algebra $C_o(S)$ has a sequential (or countable) approximate identity if and only if S is σ-compact.

<u>Proof</u>. Assume S is σ-compact, then S can be represented as $S = \bigcup_{n=1}^{\infty} U_n$, where each U_n is a relatively compact open set, and $\overline{U}_n \subset U_{n+1}$ for all n. Since S is locally compact there exist real-valued continuous functions e_n on S such that $0 \leq e_n \leq 1$, $e_n(s) = 1$ for $s \in \overline{U}_n$, and $e_n(s) = 0$ for $s \in S \setminus U_{n+1}$. Since the support of e_n is contained in the compact set $\overline{U}_{n+1}$, we have $e_n \in C_o(S)$. Clearly, $\{e_n\}_{n=1}^{\infty}$ is a sequential (countable) approximate identity in $C_o(S)$.

Conversely, assume $C_o(S)$ has a sequential approximate identity $\{e_n\}_{n=1}^{\infty}$. Since $e_n \in C_o(S)$ there is a compact set K_n such that $|e_i(s)| < \frac{1}{n}$ $(1 \leq i \leq n)$ for all $s \in S \setminus K_n$. Then all e_n vanish except in $\bigcup_{n=1}^{\infty} K_n$. Now fix an $s_o \in S$. Then there exists a function $f \in C_o(S)$ with $f(s_o) = 1$. Since $\{e_n\}_{n=1}^{\infty}$ is an approximate identity in $C_o(S)$ it follows that

$$1 = f(s_o) = \lim_n e_n(s_o)f(s_o) = \lim_n e_n(s_o).$$

Thus $e_n(s_o) \neq 0$ for some n and so $s_o \in \bigcup_{n=1}^{\infty} K_n$. Since $s_o \in S$ was arbitrary, $S = \bigcup_{n=1}^{\infty} K_n$; that is, S is σ-compact. The countable case is similar and left to the reader. □

§12. C^*-ALGEBRAS

Examples of noncommutative C^*-algebras are provided by norm closed *-subalgebras of the algebra $B(H)$ of bounded linear operators on a Hilbert space H. Multiplication in $B(H)$ is operator composition, the involution $T \to T^*$ is the usual adjoint operation, and the norm is the operator norm

$$||T|| = \sup\{||T\xi||: \xi \in H, ||\xi|| \leq 1\}.$$

By the Gelfand-Naimark theorem every C^*-algebra is isometrically *-isomorphic to a norm closed *-subalgebra of bounded linear operators on some Hilbert space (see Dixmier [70, p. 45]).

Let A be a C^*-algebra. An element $x \in A$ is called <u>positive</u>, and we write $x \geq 0$, if x is hermitian (that is, $x^* = x$) and if the spectrum of x in A is a subset of the nonnegative reals. It is well known that $x^*x \geq 0$ for all $x \in A$, and that the set of positive elements in A forms a closed convex cone [70, (1.6.1), p. 15]. A <u>positive functional</u> p on A is a linear functional on A such that $p(x^*x) \geq 0$ for all $x \in A$. One can show that an element $x \in A$ is positive if and only if $p(x) \geq 0$ for all positive functionals p on A [70, (2.6.2), p. 46].

(12.3) <u>Definition</u>. We say that an approximate identity $\{e_\lambda\}_{\lambda\in\Lambda}$ in a C^*-algebra is <u>increasing</u> if $e_\lambda \geq 0$ and if $\lambda_1 \leq \lambda_2$ implies that $e_{\lambda_1} \leq e_{\lambda_2}$ for all $\lambda_1, \lambda_2 \in \Lambda$.

(12.4) <u>Theorem</u>. Every C^*-algebra has an increasing approximate identity bounded by one.

Proof. Let A be a C^*-algebra and extend the C^*-norm to A_1 by

$$||x + \alpha|| = \sup\{||xy + \alpha y||: y \in A, ||y|| \leq 1\}.$$

Then A_1 is a C^*-algebra with identity [70, (1.3.8), p. 10]. For any $\lambda = \{x_1,\ldots,x_n\}$ in the set Λ of all finite subsets of A ordered by inclusion set

$$h_\lambda = x_1^*x_1 + \cdots + x_n^*x_n.$$

Then $h_\lambda \geq 0$ and so

$$e_\lambda = nh_\lambda(1 + nh_\lambda)^{-1}$$

is a well defined hermitian element in A. By the functional representation of h_λ as a nonnegative function we see that $e_\lambda \geq 0$ and $||e_\lambda|| \leq 1$.

We want to show that $\{e_\lambda\}_{\lambda\in\Lambda}$ is an approximate identity. Observe that

$$[x_i(1 - e_\lambda)]^*[x_i(1 - e_\lambda)] \leq \sum_{j=1}^{n} [x_j(1 - e_\lambda)]^*[x_j(1 - e_\lambda)]$$

$$= (1 - e_\lambda)h_\lambda(1 - e_\lambda)$$

$$= h_\lambda(1 + nh_\lambda)^{-2} \leq \frac{1}{4n};$$

where the last inequality follows from the fact that the real function $t \to t(1 + nt)^{-2}$, $(t \geq 0)$, has maximum value $\frac{1}{4n}$. Thus

$$||x_i(1 - e_\lambda)||^2 = ||[x_i(1 - e_\lambda)]^*[x_i(1 - e_\lambda)]|| \leq \frac{1}{4n}.$$

Now for arbitrary $x \in A$ and $\varepsilon > 0$ choose a finite set λ_o of n

elements in A such that $x \in \lambda_o$ and $n > \varepsilon^{-2}$. Then for all $\lambda \geq \lambda_o$,

$$||x - xe_\lambda|| = ||x(1 - e_\lambda)|| < \varepsilon.$$

Hence

$$\lim xe_\lambda = x \quad \text{for every} \quad x \in A;$$

and by the continuity of the involution also

$$\lim e_\lambda x = (\lim x^* e_\lambda)^* = (x^*)^* = x.$$

Thus $\{e_\lambda\}_{\lambda \in \Lambda}$ is an approximate identity in A.

It only remains to show that $\lambda_1 \leq \lambda_2$ implies $e_{\lambda_1} \leq e_{\lambda_2}$. We have $\lambda_1 = \{x_1,\ldots,x_n\}$, $\lambda_2 = \{x_1,\ldots,x_m\}$ with $n \leq m$. Then

$$0 \leq h_{\lambda_1} \leq h_{\lambda_2},$$

and so [70, (1.6.8), p. 18]

$$(1 + mh_{\lambda_2})^{-1} \leq (1 + mh_{\lambda_1})^{-1} \leq (1 + nh_{\lambda_1})^{-1}.$$

Finally,

$$e_{\lambda_1} = 1 - (1 + nh_{\lambda_1})^{-1} \leq 1 - (1 + mh_{\lambda_2})^{-1} = e_{\lambda_2};$$

thus $\{e_\lambda\}_{\lambda \in \Lambda}$ is increasing.□

(12.5) Remark. Every separable C^*-algebra A has a sequential increasing approximate identity bounded by one. Indeed, since A is separable there exists a sequence $\{x_n\}_{n=1}^{\infty}$ of elements in A which is dense in A. Set $e_n = e_{\{x_1,\ldots,x_n\}}$ which is defined as in the proof of Theorem (12.4). Then $\{e_n\}_{n=1}^{\infty}$ is a sequential increasing approximate identity bounded by one for a dense subset of A and so for all of A. We will improve this result by showing that every separable C^*-algebra has a sequential approximate identity of the type $C_o(S)$ has for σ-compact S.

An element x in a C^*-algebra A is said to be <u>strictly positive</u> if $p(x) > 0$ for each nonzero positive functional p on A. Clearly, a strictly positive element is positive.

If x is a strictly positive element in a C^*-algebra A, and π is a nondegenerate *-representation of A on a Hilbert space H, then $\overline{\pi(x)H} = H$. Indeed, suppose $0 \neq \xi \in \overline{\pi(x)H}^{\perp}$. Since π is nondegenerate there is an element $a \in A$ such that $\pi(a)\xi \neq 0$. Define a positive functional p on A by $p(a) = (\pi(a)\xi|\xi)$. Then $p(a^*a) = (\pi(a^*a)\xi|\xi) = (\pi(a)\xi|\pi(a)\xi) = ||\pi(a)\xi||^2 \neq 0$; thus p is nonzero. On the other hand $p(x) = (\pi(x)\xi|\xi) = 0$ which contradicts the assumption that x is strictly positive. Hence $\overline{\pi(x)H}^{\perp} = \{0\}$ and so $\overline{\pi(x)H} = H$.

(12.6) <u>Lemma</u>. Let A be a normed algebra with a sequential left approximate identity $\{f_k\}_{k=1}^{\infty}$. Then any bounded left approximate identity $\{e_\lambda\}_{\lambda\in\Lambda}$ of A contains a sequential bounded left approximate identity $\{e_{\lambda_n}\}_{n=1}^{\infty}$.

<u>Proof</u>. Choose $\lambda_n \in \Lambda$ inductively such that $\lambda_n \geq \lambda_{n-1}$ and

$$||f_k - e_{\lambda_n} f_k|| \leq \frac{1}{n} \quad (1 \leq k \leq n).$$

Then $\{e_{\lambda_n}\}_{n=1}^{\infty}$ is a sequential bounded left approximate identity in A. Indeed, let $x \in A$ and choose any $\varepsilon > 0$. Then there exists a k_o such that

$$||x - f_{k_o} x|| \leq \varepsilon.$$

Now choose n_o such that

$$||f_{k_o} - e_{\lambda_n} f_{k_o}|| \leq \varepsilon \quad \text{for all} \quad n \geq n_o.$$

Then

$$||x - e_{\lambda_n} x|| \leq ||x - f_{k_o} x|| + ||f_{k_o} x - e_{\lambda_n} f_{k_o} x|| + ||e_{\lambda_n} x - e_{\lambda_n} f_{k_o} x||$$

$$\leq ||x - f_{k_o} x|| + ||f_{k_o} - e_{\lambda_n} f_{k_o}||\cdot||x|| + ||e_{\lambda_n}||\cdot||x - f_{k_o} x||$$

$$\leq (1 + ||x|| + K)\varepsilon$$

for all $n \geq n_o$, where K is a bound of $\{e_\lambda\}_{\lambda\in\Lambda}$. Thus $\lim e_{\lambda_n} x = x$.□

(12.7) Proposition. A C^*-algebra with a sequential left approximate identity contains a strictly positive element.

Proof. Let A be a C^*-algebra with a sequential left approximate identity. Then the increasing approximate identity of A bounded by one of Theorem (12.4) contains, by Lemma (12.6), a sequential increasing approximate identity $\{e_n\}_{n=1}^{\infty}$ bounded by one. Set

$$x = \sum_{n=1}^{\infty} 2^{-n} e_n .$$

If p is a nonzero positive functional on A, then $||p|| = \lim p(e_n)$ [70, (2.1.5)(v), p. 28]. Hence $p(e_n) > 0$ for some n, so $p(x) = \sum_1^{\infty} 2^{-n} p(e_n) > 0$. This shows that x is strictly positive, and the proof is complete.□

(12.8) Proposition. Every separable C^*-algebra contains a strictly positive element.

Proof. Let A be a separable C^*-algebra and $\{y_n\}_{n=1}^{\infty}$ dense in the unit sphere of A. Set $x_n = y_n^* y_n$; then $\{x_n\}_{n=1}^{\infty}$ is dense in the unit sphere of the set of all positive elements of A. Clearly

$$x = \sum_{n=1}^{\infty} 2^{-n} x_n$$

is strictly positive in A (alternatively, cite (12.5) and (12.7).) □

(12.9) Lemma. A C^*-algebra with a strictly positive element has a sequential increasing abelian approximate identity bounded by one.

Proof. Assume A is a C^*-algebra with a strictly positive element x_o; we may take x_o with norm equal to one. Set

$$e_n = x_o^{1/n}, \quad n = 1,2,\ldots,$$

and observe that $e_n \geq 0$, $||e_n|| = 1$, $n \leq m$ implies $e_n \leq e_m$, and e_n

and e_m commute for all n, m. We want to show that $\lim xe_n = x$ for every $x \varepsilon A$. Since $x = (x_1 - x_2) + i(x_3 - x_4)$ with $x_1, x_2, x_3, x_4 \geq 0$, it is sufficient to do this for $x \geq 0$.

Let $x \geq 0$ and set $y = x^{\frac{1}{2}}$. Since $0 \leq e_n \leq e_m \leq 1$ for all n, m with $n \leq m$, we have $0 \leq ye_ny \leq ye_my \leq y^2 = x$ [70, (1.6.8), p. 18]. Hence $\{z_n\}_{n=1}^{\infty}$, where $z_n = x - ye_ny$, is a monotone decreasing sequence of positive elements in A. We claim that $\lim ||z_n|| = 0$. Let P be the set of all positive functionals p on A with $||p|| \leq 1$. Then P is compact in the w^*-topology. We may regard each z_n as a continuous function on P by the evaluation map. Since $z_n \geq 0$, $||z_n|| = \sup\{p(z_n)\colon p \varepsilon P\}$; hence it suffices to show that z_n converges uniformly to 0 on P. As the sequence $\{z_n\}_{n=1}^{\infty}$ is monotone, this will follow from Dini's theorem once we know that $\lim p(z_n) = 0$ for each $p \varepsilon P$.

Let π be a nondegenerate $*$-representation of A on a Hilbert space H. By Spectral Theory $\pi(e_n) = \pi(x_o)^{1/n}$ converges strongly to the range projection of $\pi(x_o)$. Since x_o is strictly positive, $\overline{\pi(x_o)H} = H$; hence it follows that $\lim \pi(e_n) = I$ strongly on H, where I is the identity operator on H.

Let $p \neq 0$ be an arbitrary element of P and π_p be the associated $*$-representation of A on the Hilbert space H_p. Then π_p is non-degenerate with a cyclic vector ξ_p and

$$p(z_n) = (\pi_p(z_n)\xi_p | \xi_p)$$

$$= (\pi_p(y - e_ny)\xi_p | \pi_p(y)\xi_p)$$

$$= (\pi_p(y)\xi_p - \pi_p(e_n)\pi_p(y)\xi_p | \pi_p(y)\xi_p),$$

which converges to zero since $\lim \pi_p(e_n) = I$ strongly. Thus $\lim ||z_n|| = 0$.

Let $(1 - e_n)^{\frac{1}{2}}$ be the unique positive square root of $1 - e_n$ in A_1. Then

$$||x - xe_n||^2 = ||yy(1 - e_n)^{\frac{1}{2}}(1 - e_n)^{\frac{1}{2}}||^2$$

$$\leq ||y||^2||y(1 - e_n)^{\frac{1}{2}}||^2||(1 - e_n)||$$

$$\leq ||y||^2||y(1 - e_n)y|| = ||y||^2||z_n||,$$

and so $\lim xe_n = x$. Thus $\{e_n\}$ is an approximate identity in A. □

(12.10) Remark. Let S be a locally compact Hausdorff space and $C_o(S)$ the commutative C^*-algebra of all continuous complex-valued functions on S vanishing at infinity. Then the following conditions are equivalent:

i) S is σ-compact;

ii) $C_o(S)$ contains a function f with $f(s) > 0$ $(s \in S)$;

iii) $C_o(S)$ contains a strictly positive element;

iv) $C_o(S)$ has a sequential approximate identity.

Evidently S may be σ-compact without having a countable base for its topology, so $C_o(S)$ may have a strictly positive element without being separable.

The set R of real numbers with the discrete topology is a locally compact Hausdorff space which is not σ-compact. Thus $C_o(R)$, when R

is given the discrete topology, is an example of a commutative C^*-algebra which has no strictly positive element and no sequential approximate identity.

(12.11) Lemma. Let A be a normed algebra. If $\{e_\lambda\}_{\lambda\varepsilon\Lambda}$ is a left approximate identity in A and $\{f_\mu\}_{\mu\varepsilon M}$ is a bounded left approximate identity for $\{e_\lambda\}_{\lambda\varepsilon\Lambda}$, then $\{f_\mu\}_{\mu\varepsilon M}$ is a left approximate identity in A.

Proof. For $x \varepsilon A$,

$$||x - f_\mu x|| \leq ||x - e_\lambda x|| + ||e_\lambda x - f_\mu e_\lambda x|| + ||f_\mu x - f_\mu e_\lambda x||$$

$$\leq ||x - e_\lambda x|| + ||e_\lambda - f_\mu e_\lambda||\cdot||x|| + K||x - e_\lambda x||,$$

where K is a bound for $\{f_\mu\}_{\mu\varepsilon M}$. Choose $\varepsilon > 0$. Since $\{e_\lambda\}_{\lambda\varepsilon\Lambda}$ is a left approximate identity in A we can find a $\lambda_o \varepsilon \Lambda$ such that $||x - e_{\lambda_o} x|| < \varepsilon$. Since $\{f_\mu\}_{\mu\varepsilon M}$ is a left approximate identity for $\{e_\lambda\}_{\lambda\varepsilon\Lambda}$ we can find a $\mu_o \varepsilon M$ such that $||e_{\lambda_o} - f_\mu e_{\lambda_o}|| < \varepsilon$ for all $\mu \geq \mu_o$. Then

$$||x - f_\mu x|| \leq (1 + ||x|| + K)\varepsilon \quad \text{for all} \quad \mu \geq \mu_o,$$

and so $\lim f_\mu x = x$ for all $x \varepsilon A$. Thus $\{f_\mu\}_{\mu\varepsilon M}$ is a left approximate identity in A. □

(12.12) Theorem. Let A be a C^*-algebra with a strictly positive element x_o. If B is any closed commutative *-subalgebra of A containing x_o then B is isometrically *-isomorphic to $C_o(S)$ for some σ-compact space S. Furthermore, every bounded approximate identity in B is an approximate identity in A.

Proof. Let B be a closed commutative *-subalgebra of A containing x_0. By the Gelfand-Naimark theorem B is isometrically *-isomorphic to $C_0(S)$ for some locally compact Hausdorff space S. Since by Lemma (12.9) $e_n = x_0^{1/n}$, $n = 1,2,\ldots$, is a sequential approximate identity in B, S is σ-compact by Proposition (12.2).

Now let $\{f_\mu\}_{\mu\in M}$ be any bounded approximate identity in B. Then $\{f_\mu\}_{\mu\in M}$ is a bounded approximate identity for $\{e_n\}_{n=1}^\infty$. Hence by Lemma (12.11) $\{f_\mu\}_{\mu\in M}$ is an approximate identity in A. □

(12.13) Remark. Let A be a separable C^*-algebra. By Proposition (12.8) A contains a strictly positive element. Hence by Theorem (12.12) A has a sequential approximate identity of the type $C_0(S)$ has for σ-compact S. Here is an interesting short proof of this fact:

Let $\{x_k\}$ be a countable dense set in the hermitian part of the unit sphere of A, and set

$$x = \Sigma\, 2^{-k} x_k^* x_k .$$

Since x is a positive element of A, the C^*-algebra B generated by x is commutative and so isometrically *-isomorphic to the algebra $C_0(S)$. Since $C_0(S)$ is generated by a single function, S is σ-compact. Let $\{e_n\}$ be a bounded sequential increasing approximate identity in B as constructed in the proof of Proposition (12.2). It remains only to show that $\{e_n\}$ is an approximate identity for A. Since

$$[(1 - e_n)x_k]^*[(1 - e_n)x_k] \leq 2^k(1 - e_n)x(1 - e_n),$$

we have that

$$||x_k(1 - e_n)||^2 = ||(1 - e_n)x_k||^2$$

$$= ||[(1 - e_n)x_k]^*[(1 - e_n)x_k]|| \leq 2^k||(1 - e_n)x(1 - e_n)||,$$

so $\lim_n ||x_k(1 - e_n)|| = 0 = \lim_n ||(1 - e_n)x_k||$. Thus $\{e_n\}$ is a bounded approximate identity for the dense subset $\{x_k\}$ in the hermitian part of the unit ball of A; hence $\{e_n\}$ is an approximate identity for A.□

(12.14) Example. There is a (necessarily nonseparable) C^*-algebra A with no bounded abelian approximate identity. In particular, no maximal commutative *-subalgebra of A contains an approximate identity for A.

Proof. Let Γ be an uncountable set and let H be the Hilbert space direct sum of $\Gamma \times \Gamma$ copies of the two-dimensional Hilbert space C^2; that is, $H = \{\xi = (\xi_{\alpha\beta}): \xi_{\alpha\beta} \in C^2$ and $\sum_{\alpha,\beta\in\Gamma} ||\xi_{\alpha\beta}||^2 < \infty\}$. Any matrix $T = (t_{\alpha\beta})$ with $t_{\alpha\beta} \in L(C^2)$ and $\sup\{||t_{\alpha\beta}||: \alpha,\beta \in \Gamma\} < \infty$ defines a bounded linear operator on H by $(T\xi)_{\alpha\beta} = t_{\alpha\beta}\xi_{\alpha\beta}$. Let p and q be noncommuting orthogonal projections on C^2. For each $\gamma \in \Gamma$, define a bounded linear operator p_γ on H by

$$(p_\gamma)_{\gamma\gamma} = I \text{ (the identity operator on } C^2),$$

$$(p_\gamma)_{\gamma\beta} = p \quad \text{if } \gamma \neq \beta,$$

$$(p_\gamma)_{\alpha\gamma} = q \quad \text{if } \alpha \neq \gamma,$$

$$(p_\gamma)_{\alpha\beta} = 0 \quad \text{if } \alpha \neq \gamma \text{ and } \beta \neq \gamma.$$

Let A be the C^*-algebra generated by $\{p_\gamma\}_{\gamma\varepsilon\Gamma}$. If $\alpha \neq \gamma$, p_α and hence $p_\gamma p_\alpha$ and $p_\alpha p_\gamma$ have only one nonzero entry in the γ-th row and only one in the γ-th column, neither being on the diagonal. Since any polynomial f in the p_α with zero constant term is of the form "λp_γ + terms containing p_α for $\alpha \neq \gamma$", we see that $f - \lambda p_\gamma$ has a zero entry in the $\gamma\gamma$-th place and only a finite number of nonzero entries in both the γ-th row and γ-th column. The scalar λ is given by $f_{\gamma\gamma} = \lambda I$.

Suppose that B is a commuting subset of A containing left approximate units for A bounded by K. Let $\varepsilon > 0$ be such that $(1 - 2\varepsilon)^2||pq - qp|| > 4K\varepsilon$. Then, for each $\gamma \varepsilon \Gamma$, there is an element $u_\gamma \varepsilon B$ with $||u_\gamma|| \leq K$ and $||p_\gamma - u_\gamma p_\gamma|| < \varepsilon$. Let f_γ be a polynomial in the p_α with zero constant term, $||f_\gamma|| \leq K$, and $||u_\gamma - f_\gamma|| < \varepsilon$. If λ_γ is the coefficient of p_γ in f_γ, then

$$|1 - \lambda_\gamma| = ||I - (f_\gamma)_{\gamma\gamma}|| \leq ||I - (u_\gamma)_{\gamma\gamma}I|| + ||(u_\gamma)_{\gamma\gamma} - (f_\gamma)_{\gamma\gamma}||$$

$$\leq ||p_\gamma - u_\gamma p_\gamma|| + ||u_\gamma - f_\gamma|| < 2\varepsilon.$$

Since Γ is uncountable, there exists an integer n such that the set $\Gamma_n = \{\gamma \varepsilon \Gamma\colon f_\gamma - \lambda_\gamma p_\gamma$ has at most n nonzero entries in both the γ-th row and column$\}$ is infinite. Let $\gamma_0,\ldots,\gamma_n$ be n+1 distinct elements in Γ_n. Choose $\gamma \varepsilon \Gamma_n \setminus \{\gamma_0,\ldots,\gamma_n\}$ such that $(f_{\gamma_i} - \lambda_{\gamma_i} p_{\gamma_i})_{\gamma_i\gamma} = 0 = (f_{\gamma_i} - \lambda_{\gamma_i} p_{\gamma_i})_{\gamma\gamma_i}$ for all i, $0 \leq i \leq n$. We have $(f_\gamma - \lambda_\gamma p_\gamma)_{\gamma_i\gamma} = 0$ for some value of i because the equality fails for at most n values of i. Hence

$$0 = ||u_{\gamma_i} u_\gamma - u_\gamma u_{\gamma_i}|| \geq ||f_{\gamma_i} f_\gamma - f_\gamma f_{\gamma_i}|| - 4K\varepsilon$$

$$\geq ||(f_{\gamma_i} f_\gamma - f_\gamma f_{\gamma_i})_{\gamma_i \gamma}|| - 4K\varepsilon$$

$$= |\lambda_\gamma| \cdot |\lambda_{\gamma_i}| \cdot ||pq - qp|| - 4K\varepsilon$$

$$> (1 - 2\varepsilon)^2 ||pq - qp|| - 4K\varepsilon > 0.$$

This contradiction shows that the C^*-algebra A has no bounded abelian approximate identity. □

13. Group algebras.

In this section we briefly study approximate identities in the group algebra $L^1(G)$ of a locally compact topological group G. A substantial literature exists concerning such approximate identities and we refer the reader to the notes at the end of this monograph for information which goes beyond what is given here.

Let G be a locally compact (Hausdorff) topological group and let $C_{oo}(G)$ be the space of all complex-valued continuous functions on G with compact support. Let μ be the left invariant Haar measure on G. Then it is straightforward to verify that

$$||f||_p = (\int_G |f|^p \, d\mu)^{1/p}, \quad f \in C_{oo}(G),$$

defines a norm in the linear space $C_{oo}(G)$ for each p, $1 \leq p < \infty$. $L^p(G)$ is the Banach space obtained by completing $C_{oo}(G)$ with respect to this norm. The space $L^1(G)$ is a Banach algebra under convolution multiplication defined by

$$(f*g)(x) = \int_G f(xy)g(y^{-1}) \, d\mu(y) = \int_G f(y)g(y^{-1}x) \, d\mu(y),$$

for $f, g \in L^1(G)$. The group algebra is commutative if and only if G is commutative. Further, $L^1(G)$ becomes a Banach *-algebra with respect to the (isometric) involution defined by $f^*(x) = \Delta(x^{-1})\overline{f(x^{-1})}$, $x \in G$, where Δ is the modular function of G.

If $f \in L^p(G)$, let f_a and ${}_af$ denote the translation functions defined on G by $f_a(x) = f(xa)$ and ${}_af(x) = f(ax)$. Then from the left invariance of μ, $||f_a||_p = \Delta(a)^{-1/p}||f||_p$ and $||{}_af||_p = ||f||_p$. We remark that the spaces $L^p(G)$, for $p > 1$, are in general not closed with respect to convolution.

(13.1) Theorem. The group algebra $L^1(G)$ has a left (or right) identity if and only if G is discrete.

Proof. If G is discrete, the points of G are congruent open sets having equal positive Haar measure which may be taken as 1. Then f is summable if and only if $f(x) = 0$ except on a countable set $\{x_n\}$ and $\Sigma\, |f(x_n)| < \infty$. The function e which is 1 at $x = 1$ and zero elsewhere is an identity:

$$(f*e)(x) = \int_G f(y)e(y^{-1}x)\, d\mu(y) = \sum_{y\in G} f(y)e(y^{-1}x) = f(x).$$

Conversely suppose that e is an identity for $L^1(G)$. We show that there is a positive lower bound to the measures of nonempty open Baire sets. Otherwise, given any $\varepsilon > 0$ there exists an open neighborhood V of the identity in G whose measure is less than ε, and hence one such that $\int_V |e|\, d\mu < \varepsilon$. Choose a symmetric U such that $U^2 \subset V$ and let χ_U be its characteristic function. Then

$$\chi_U(x) = (e*\chi_U)(x) = \int_G e(y)\chi_U(y^{-1}x)\, d\mu(y) = \int_{xU} e(y)\, d\mu(y) \leq \int_V |e|\, d\mu < \varepsilon$$

for almost all x in U, contradicting $\chi_U(x) \equiv 1$ in U. Therefore, there is a number $\alpha > 0$ such that the measure of every nonempty open Baire set is at least α. From this it follows at once that every open set whose closure is compact, and which therefore has finite measure, contains only a finite set of points, since otherwise its measure is seen to be $\geq n\alpha$ for every n by choosing n disjoint nonempty open subsets. Therefore, every point is an open set, and the topology is discrete. □

Although $L^1(G)$ does not ordinarily possess an identity element, it always does contain an approximate identity. This will be a consequence of the following two lemmas whose proofs we include for completeness.

(13.2) Lemma. If $f \in L^p(G)$, $1 \leq p < \infty$, then the map $x \to {}_xf$ of G into $L^p(G)$ is right uniformly continuous.

Proof. Let $\varepsilon > 0$ and choose a continuous function g on G with compact support C such that $||f - g||_p < \frac{\varepsilon}{3}$. Fix a compact symmetric neighborhood U of the identity e in G. Using the standard fact that g is right uniformly continuous on G there is a symmetric neighborhood V of e contained in U such that $|g(s) - g(t)| < \frac{\varepsilon}{3}\cdot\mu(UC)^{-1/p}$ for all $s, t \in G$ such that $st^{-1} \in V$; thus

$$||g - {}_xg||_\infty < \frac{\varepsilon}{3}\cdot\mu(UC)^{-1/p}$$

for all $x \in V$. Hence $||g - {}_xg||_p < \frac{\varepsilon}{3}$ for all $x \in V$ and so

$$||f - {}_xf||_p \leq ||f - g||_p + ||g - {}_xg||_p + ||{}_xg - {}_xf||_p < \varepsilon$$

for all $x \in V$. Then for $x = st^{-1} \in V$ we have $||{}_sf - {}_tf||_p = ||{}_xf - f||_p < \varepsilon$ as required. □

(13.3) Lemma. Given $f \in L^p(G)$, $1 \leq p < \infty$, and $\varepsilon > 0$, there exists a neighborhood V of the identity in G such that

$$||f - u*f||_p < \varepsilon \quad \text{and} \quad ||f - f*u||_p < \varepsilon$$

whenever u is any nonnegative real-valued function in $L^1(G)$ such that $u(x) = 0$ for $x \notin V$ and $\int_G u \, d\mu = 1$.

Proof. If f and g are functions on G such that $f\overline{g} \in L^1(G)$, we write $(f,g) = \int_G f\overline{g} \, d\mu$. Let $g \in L^q(G)$, where $\frac{1}{p} + \frac{1}{q} = 1$. If u is any nonnegative function in $L^1(G)$, then $u*f \in L^p(G)$ and hence, by Hölder's inequality, $(u*f - f)\overline{g} \in L^1(G)$, so

$$\begin{aligned}(u*f - f,g) &= \int_G [(u*f)(x) - f(x)]\overline{g(x)} \, d\mu(x) \\ &= \int_G \int_G u(y)[f(y^{-1}x) - f(x)]\overline{g(x)} \, d\mu(y)d\mu(x).\end{aligned}$$

Interchanging the order of integration (Fubini) and applying Hölder's inequality gives

$$|(u*f - f, g)| \leq ||g||_q \int_G ||{}_{y^{-1}}f - f||_p u(y) \, d\mu(y).$$

Then, considering the operator norm of the linear functional $g \to (u*f - f,g)$ on $L^q(G)$, we have

$$||u*f - f||_p \leq \int_G ||{}_{y^{-1}}f - f||_p u(y) \, d\mu(y).$$

By (13.2), there is a neighborhood V of e such that $||{}_{y^{-1}}f - f||_p < \varepsilon$ for all $y \in V$. Therefore if $\int_G u(x) \, d\mu(x) = 1$ and $u(y) = 0$ for $y \notin V$, then $||u*f - f||_p < \varepsilon$.

To prove the other inequality, recall that the modular function Δ is a continuous homomorphism of G into $(0,\infty)$, $\Delta(e) = 1$, and

$$\int_G u(x^{-1})\Delta(x^{-1})\, d\mu(x) = \int_G u(x)\, d\mu(x)$$

for any $u \in L^1(G)$. Now set $m = \int_G u(x^{-1})\, d\mu(x)$ (we see below that m is finite and nonzero), and let $g \in L^q(G)$. Then, as above,

$$|(f*u - f, g)| = \left|\int_G\int_G [f(xy) - f(x)/m]u(y^{-1})\overline{g(x)}\, d\mu(y)d\mu(x)\right|$$

$$\leq ||g||_q \int_G ||mf_y - f||_p[u(y^{-1})/m]\, d\mu(y).$$

Thus, $$||f*u - f||_p \leq \int_G ||mf_y - f||_p[u(y^{-1})/m]\, d\mu(y).$$

Observe that $m \to 1$ as the neighborhoods V of e decrease (indeed, since $m = \int_G u(x^{-1})\Delta(x^{-1})\Delta(x)d\mu(x)$, then $a = a \int_G u(x)d\mu(x) = a \int_G u(x^{-1})\Delta(x^{-1})d\mu(x) \leq m \leq A \int_G u(x^{-1})\Delta(x^{-1})d\mu(x) = A \int_G u(x)d\mu(x) = A$, where $a = \min_{x\in V} \Delta(x)$ and $A = \max_{x\in V} \Delta(x)$.) Hence, there is a neighborhood V of e with $||mf_y - f||_p < \varepsilon$ if $y \in V$, and then $||f*u - f||_p < \varepsilon \int_G [u(y^{-1})/m]\, d\mu(y) = \varepsilon$. □

(13.4) <u>Theorem</u>. The group algebra $L^1(G)$ has a two-sided approximate identity bounded by 1.

<u>Proof</u>. The neighborhoods V of the identity in G form a directed set under inclusion, and if u_V is a nonnegative real-valued function vanishing off V and satisfying $\int_G u_V\, d\mu = 1$, then u_V*f and $f*u_V$ converge to f in the L^p-norm for any $f \in L^p(G)$ by (13.3). □

(13.5) Definition. Let A be a normed algebra. An approximate identity $\{e_\lambda\}_{\lambda\in\Lambda}$ in A is called central if $e_\lambda x = xe_\lambda$ for all $x \in A$, $\lambda \in \Lambda$.

The following stronger version of (13.4) holds for compact groups.

(13.6) Theorem. The group algebra $L^1(G)$ of a compact group G has a central approximate identity bounded by 1.

Proof. Take any neighborhood basis $\{U_\lambda\}_{\lambda\in\Lambda}$ of the identity in G such that $U_\lambda = U_\lambda^{-1}$. Set $V_\lambda = \cap\{x^{-1}U_\lambda x: x \in G\}$. Since G is compact, $\{V_\lambda\}_{\lambda\in\Lambda}$ is a neighborhood basis of the identity in G such that

$$x^{-1}V_\lambda x = V_\lambda = V_\lambda^{-1} \quad \text{for all } x \in G,\ \lambda \in \Lambda.$$

Take

$$e_\lambda = \left(\int_{V_\lambda} d\mu\right)^{-1}\chi_{V_\lambda},$$

where χ_{V_λ} is the characteristic function of V_λ. It is easy to check that e_λ is central in $L^1(G)$; so $\{e_\lambda\}_{\lambda\in\Lambda}$ is the desired approximate identity. □

Remark. Those topological groups whose group algebras have central approximate identities have been characterized. A locally compact group G is called an [SIN]-group if it contains a fundamental family of compact invariant neighborhoods of the identity. Then, a locally compact group G is an [SIN]-group if and only if $L^1(G)$ has a central approximate identity (see Section 13 of the Notes at the end of this volume for more on this.)

Now consider the special Hilbert space $L^2(G)$. There exists a natural

*-representation of $L^1(G)$ on $L^2(G)$ which may be defined as follows: For $f \in L^1(G)$ and $g \in C_{oo}(G)$ it can be shown that $f*g \in L^2(G)$ and furthermore $||f*g||_2 \le ||f||_1||g||_2$. Hence the map $f \to f*g$ is continuous relative to the L^2-norm. Since $C_{oo}(G)$ is dense in $L^2(G)$, it follows that there is a well-defined bounded linear operator T_f on $L^2(G)$ such that $T_f g = f*g$ for $g \in C_{oo}(G)$. The map $f \to T_f$ is an injective *-representation of $L^1(G)$ on $L^2(G)$.

Let $C^*_r(G)$ denote the C^*-algebra obtained by completing $L^1(G)$ in the C^*-norm $f \to ||T_f||$. If G is discrete, $L^1(G)$ has an identity and hence so does $C^*_r(G)$. To prove the converse we need the following lemma.

(13.7) Lemma. Let G be a nondiscrete locally compact topological group. Then, if $f \in L^1(G)$, there is a sequence $\{g_n\}$ of elements in $L^2(G)$ with $||g_n||_2 = 1$ such that $\lim_n ||T_f g_n||_2 = 0$.

Proof. It suffices to prove the result for $f \in C_{oo}(G)$, a norm-dense subset of $L^1(G)$. Suppose $f \in C_{oo}(G)$, and the support of f is K, a compactum. Let $\{V_n\}$ be a decreasing sequence of compact symmetric neighborhoods of the identity of G such that $\lim \mu(V_n) = 0$, where μ is the left Haar measure on G, and let $g_n = \chi_{V_n} / (\mu(V_n))^{1/2}$. Then

$$T_f g_n(x) = (f*g_n)(x) = \int_G f(y) g_n(y^{-1}x)\, d\mu(y) = 0 \quad \text{for all} \quad x \notin KV_n$$

and

$$|T_f g_n(x)| \le ||f||_\infty \mu(V_n)^{1/2} \quad \text{for all} \quad x \in G.$$

Hence, $||T_f g_n||_2^2 \le ||f||_\infty^2 \mu(V_n)\mu(KV_n) \to 0$ as $n \to \infty$. This completes the proof. $\square$

(13.8) Proposition. If G is a nondiscrete locally compact group, then $C_r^*(G)$ does not have an identity.

Proof. Suppose $C_r^*(G)$ does have an identity 1 and observe that $C_r^*(G)$ can be identified with the C^*-subalgebra of $B(L^2(G))$ generated by the operators T_f, $f \in L^1(G)$. Let $f \in L^1(G)$ be such that $||1 - T_f|| \leq \frac{1}{4}$ and let $g \in L^2(G)$ with $||g||_2 = 1$ be such that $||T_f g||_2 \leq \frac{1}{4}$. Then

$$||(1 - T_f)g||_2 = ||g - T_f g||_2 \geq ||g||_2 - ||T_f g||_2 \geq 1 - \frac{1}{4} = \frac{3}{4},$$

which is a contradiction. □

The group C^*-algebra $C^*(G)$ of G is obtained by completing $L^1(G)$ with respect to the C^*-norm $f \to \sup ||\pi_f||$, where π ranges over all $*$-representations of $L^1(G)$ on a Hilbert space.

(13.9) Theorem. The group C^*-algebra $C^*(G)$ has an identity if and only if G is discrete.

Proof. If G is discrete, $L^1(G)$ has an identity and hence so does $C^*(G)$. On the other hand if G is nondiscrete then $C_r^*(G)$ has no identity and since $C_r^*(G)$ is a homomorphic image of $C^*(G)$, then $C^*(G)$ also has no identity. □

Remark. Note that (13.1) may be viewed as a corollary of (13.9).

CHAPTER II

FACTORIZATION IN BANACH MODULES

If A is a Banach algebra and $z \in A$, when can z be "factored" into a product $z = xy$, with $x, y \in A$? Clearly this can be done when A has an identity. What can be said when an identity element is not present? In this chapter we shall thoroughly examine this and related questions. After preliminaries have been handled, we present our main factorization theorem in Section 16. In terms of a Banach algebra A with bounded left approximate identity the theorem asserts that for z in A there are elements x, y in A such that $z = xy$, y belongs to the closed left ideal generated by z, and y is arbitrarily close to z.

This theorem, due to Paul Cohen [51], generalized earlier factorization theorems for specific algebras due to R. Salem [240] and Walter Rudin [234], [235]. Cohen observed that the essential ingredient in one of Rudin's arguments was the presence of an approximate identity, the Fejér kernel, in the algebra $L^1(R)$.

The basic idea introduced by Cohen in his proof has been used by all subsequent writers on factorization. His argument in brief is as follows: one adjoins an identity to the algebra A, and then constructs, using the approximate identity, a sequence $\{a_n\}$ of invertible elements in A_1 such that $\{a_n\}$ converges to an element in A and $\{a_n^{-1}z\}$ also converges, even though $\{a_n^{-1}\}$ is an unbounded sequence in A. The elements x and y are then obtained from these convergent sequences.

In the remainder of the chapter many refinements, extensions, and variations of Cohen's original factorization theorem are presented along with several applications. We refer the reader to the individual sections for a description of what takes place there. Examples of algebras which do not factor are considered in Section 23.

14. Banach modules.

(14.1) Definition. Let A be a Banach algebra. A *left normed A-module* is a normed space X together with a map

$$A \times X \to X$$
$$(a,x) \to ax$$

having the following properties:

i) $(a + b)x = ax + bx$ and $a(x + y) = ax + ay$;

ii) $(\alpha a)x = \alpha(ax) = a(\alpha x)$;

iii) $(ab)x = a(bx)$;

iv) $||ax|| \leq M \cdot ||a|| \cdot ||x||$

for all $a, b \in A$, $x, y \in X$, and complex α, where M is a constant ≥ 1. *Right normed A-modules* are defined similarly. The space X is called a *normed A-bimodule* if it is both a normed left and right A-module and the module multiplication satisfies:

v) $a(xb) = (ax)b$ for all $a, b \in A$ and $x \in X$.

If X is complete, then X is called a *left* (resp. *right*) *Banach A-module*.

(14.2) Remark. Clearly, if X is a left Banach A-module, condition (14.1.iv) is equivalent to the requirement that the bilinear mapping $(a,x) \to ax$ from $A \times X \to X$ be continuous.

(14.3) Remark. Let A_1 be the Banach algebra obtained from A by adjoining an identity 1 with the usual norm $||a + \alpha 1|| = ||a|| + |\alpha|$. If X is a left Banach A-module, then X is also a left Banach A_1-module where

$$(a + \alpha 1)x = ax + \alpha x \quad \text{for} \quad a + \alpha 1 \in A_1 \quad \text{and} \quad x \in X.$$

Properties (14.1.i) - (14.1.iii) are evident and (14.1.iv) holds for the same constant M because $M \geq 1$.

(14.4) Remark. If we can take $M = 1$ in (14.1.iv), then we say that X is an isometric left Banach A-module.

Observe that every left Banach A-module X can be renormed with an equivalent norm with respect to which it is an isometric left Banach A-module. For example, define a new norm $||\cdot||'$ on X by

$$||x||' = \sup\{||ax||: a \in A_1, \ ||a|| \leq 1\}.$$

Then

$$||\cdot|| \leq ||\cdot||' \leq M||\cdot||,$$

so that $||\cdot||'$ is an equivalent norm on X and it is easily seen that $||ax||' \leq ||a||\cdot||x||'$ for all $a \in A$, $x \in X$.

15. Essential Banach modules.

(15.1) Definition. Let A be a Banach algebra and let X be a left Banach A-module. Then the closed linear subspace of X spanned by

$$AX = \{ax : a \in A, x \in X\}$$

is called the <u>essential part</u> of X and is denoted by X_e. If $X = X_e$ then X is said to be an <u>essential left Banach A-module</u>.

(15.2) <u>Proposition</u>. Let A be a Banach algebra with a bounded l.a.i. $\{e_\lambda\}_{\lambda\in\Lambda}$ and let X be a left Banach A-module. Then

$$X_e = \overline{AX} = \{x \in X : \lim e_\lambda x = x\}.$$

<u>Proof</u>. Suppose $x \in X_e$. Then for any $\varepsilon > 0$ we can find $a_1,\ldots,a_n \in A$, $x_1,\ldots,x_n \in X$ such that

$$||x - \sum_{k=1}^{n} a_k x_k|| < \varepsilon.$$

Since

$$\lim_\lambda e_\lambda(\sum_{k=1}^{n} a_k x_k) = \sum_{k=1}^{n} \lim_\lambda (e_\lambda a_k)x = \sum_{k=1}^{n} a_k x_k$$

and $\{e_\lambda\}$ is bounded it follows that $\lim e_\lambda x = x$. □

(15.3) <u>Corollary</u>. Let A be a Banach algebra with a bounded l.a.i. $\{e_\lambda\}$. Then a left Banach A-module X is essential if and only if $\lim e_\lambda x = x$ for all $x \in X$.

(15.4) <u>Remark</u>. Let A be a Banach algebra with a bounded left approximate identity. Then viewed as a left module over itself, A is an essential left Banach A-module.

(15.5) Proposition. Let A be a Banach algebra with a bounded left approximate identity. If X is a left Banach A-module, then the essential part X_e of X is a submodule of X, X_e is essential, and X_e contains any submodule of X which is essential.

16. Factorization.

(16.1) Theorem. Let A be a Banach algebra with a left approximate identity bounded by $K \geq 1$ and let X be a left Banach A-module. Then for every $z \in X_e$ and $\delta > 0$ there exist elements $a \in A$ and $y \in X$ such that:

i) $z = ay$;

ii) $||a|| \leq K$;

iii) $y \in \overline{Az}$;

iv) $||y - z|| < \delta$.

In particular, $X_e = AX$.

We shall give three different proofs for this fundamental factorization theorem.

First proof. (P. J. Cohen, E. Hewitt). Fix a real number $c > 0$ such that

$$0 < K < \frac{1-c}{c};$$

for example take

$$c = \frac{1}{K+2} \quad \text{or} \quad c = \frac{1}{2K+1}.$$

For $e \in A$ with $||e|| \leq K$ define

$$\phi(e) = (1 - c + ce)^{-1}.$$

Then $\phi(e)$ is a well defined element in A_1 since

$$1 - c + ce = (1 - c)(1 + \frac{c}{1-c} e)$$

and

$$||1 - (1 + \frac{c}{1-c} e)|| = \frac{c}{1-c} ||e|| \leq \frac{c}{1-c} K < 1.$$

The inequality

$$(1 - c - cK)||\phi(e)|| \leq ||(1 - c)\phi(e)|| - ||ce\phi(e)||$$

$$\leq ||(1 - c + ce)\phi(e)|| = 1$$

shows that

$$||\phi(e)|| \leq (1 - c - cK)^{-1}.$$

Furthermore, since

$$\phi(e)x - x = \phi(e)(x - (1 - c + ce)x) = c\phi(e)(x - ex),$$

$$||\phi(e)x - x|| \leq M(1 - c - cK)^{-1}||ex - x|| \quad \text{for all} \quad x \in X.$$

We shall choose inductively a sequence $\{e_n\}$ of elements in A with $||e_n|| \leq K$, $n = 1,2,\ldots$. Set

$$a_o = 1, \quad a_n = (1 - c)^n + \sum_{k=1}^{n} c(1 - c)^{k-1}e_k, \quad n = 1,2,\ldots .$$

Then

$$\phi(e_{n+1})a_{n+1} = (1 - c)^n + \sum_{k=1}^{n} c(1 - c)^{k-1}\phi(e_{n+1})e_k,$$

$$\phi(e_{n+1})a_{n+1} - a_n = \sum_{k=1}^{n} c(1-c)^{k-1}(\phi(e_{n+1})e_k - e_k),$$

and so

$$||\phi(e_{n+1})a_{n+1} - a_n|| \le (1 - c - cK)^{-1} \sum_{k=1}^{n} ||e_{n+1}e_k - e_k||.$$

If a_n^{-1} exists and the right side of this inequality is sufficiently small, $\phi(e_{n+1})a_{n+1}$ will be invertible and then a_{n+1} will also be invertible.

Setting $$y_n = a_n^{-1}z, \quad n = 0,1,2,\ldots,$$

we have:

$$||y_{n+1} - y_n|| \le ||(\phi(e_{n+1})a_{n+1})^{-1}\phi(e_{n+1})z - a_n^{-1}\phi(e_{n+1})z||$$

$$+ ||a_n^{-1}\phi(e_{n+1})z - a_n^{-1}z||$$

$$\le M\cdot||(\phi(e_{n+1})a_{n+1})^{-1} - a_n^{-1}||\cdot||\phi(e_{n+1})||\cdot||z||$$

$$+ M\cdot||a_n^{-1}||\cdot||\phi(e_{n+1})z - z||$$

$$\le M(1 - c - cK)^{-1}||z||\cdot||(\phi(e_{n+1})a_{n+1})^{-1} - a_n^{-1}||$$

$$+ M^2(1 - c - cK)^{-1}||a_n^{-1}||\cdot||e_{n+1}z - z||.$$

Thus choosing inductively e_{n+1} in A with $||e_{n+1}|| \le K$ such that

$$||e_{n+1}e_k - e_k|| \quad (1 \le k \le n) \quad \text{and} \quad ||e_{n+1}z - z||$$

are sufficiently small, we can conclude that a_{n+1}^{-1} exists in A_1 and

$$||y_{n+1} - y_n|| < \frac{\delta}{2^{n+1}} \quad \text{for all} \quad n = 0,1,2,\dots .$$

Then

$$\sum_{n=0}^{\infty} ||y_{n+1} - y_n|| < \sum_{n=0}^{\infty} \delta/2^{n+1} = \delta.$$

It follows that $\{y_n\}$ is a Cauchy sequence in X and so has a limit $y \in \overline{Az}$ (since $y_n = a_n^{-1}z$ and $a_n^{-1} \in A_1$, then $y_n \in A_1 z$ and hence $y \in \overline{A_1 z}$. But $\overline{A_1 z} = \overline{Az} + Cz$, where $z \in \overline{Az}$ (see (15.2)), and $\overline{Az}$ is a vector space, so $y \in \overline{Az}$.) Since

$$y - z = \sum_{n=0}^{\infty} (y_{n+1} - y_n)$$

we see that

$$||y - z|| < \delta.$$

Define

$$a = \sum_{k=1}^{\infty} c(1 - c)^{k-1} e_k;$$

the series converges since $||e_k|| \leq K$. Clearly a belongs to A,

$$||a|| \leq \sum_{k=1}^{\infty} c(1 - c)^{k-1} K = K;$$

and since $a = \lim a_n$,

$$z = \lim a_n(a_n^{-1}z) = ay. \ \square$$

<u>Remark</u>. Suppose that the bounded left approximate identity in A can be chosen from a certain closed convex subset C of A. Then the factor a in $z = ay$ can be taken to be in C. Indeed,

$$a_n' = (1-c)^n e_n + \sum_{k=1}^{n} c(1-c)^{k-1} e_k$$

is a convex combination of $e_1, e_2, \ldots, e_n$, and $a = \lim a_n'$.

Second proof. (P. Koosis, M. Altman). Fix a real number $c > 0$ such that

$$0 < K < \frac{1-c}{c};$$

for example take

$$c = \frac{1}{K+2} \quad \text{or} \quad c = \frac{1}{2K+1}\,.$$

For $e \in A$ with $||e|| \leq K$ define $\phi(e)$ as before, i.e.,

$$\phi(e) = (1 - c + ce)^{-1}.$$

Then $\phi(e)$ is a well defined element in A_1 with

$$||\phi(e)|| \leq (1 - c - cK)^{-1}.$$

We shall choose inductively a sequence $\{e_n\}$ of elements in A with $||e_n|| \leq K$, $n = 1,2,\ldots$. Set

$$a_o = 1, \quad a_{n+1} = (1 - c + ce_{n+1})a_n, \quad n = 0,1,2,\ldots\,.$$

Then

$$a_n = (1-c)^n + a_n', \quad \text{where} \quad a_n' \in A,$$

and

$$a_{n+1} - a_n = (1 - c + ce_{n+1})a_n - a_n = c(e_{n+1}a_n - a_n)$$

$$= c(1 - c)^n(e_{n+1} - 1) + c(e_{n+1}a'_n - a'_n).$$

Thus the infinite series

$$\sum_{n=0}^{\infty} ||a_{n+1} - a_n||$$

converges if the series

$$\sum_{n=0}^{\infty} ||e_{n+1}a'_n - a'_n||$$

converges.

Now set

$$y_n = a_n^{-1}z, \quad n = 0,1,2,\dots .$$

then

$$y_{n+1} - y_n = (a_{n+1}^{-1} - a_n^{-1})z$$

$$= (a_{n+1}^{-1} - a_{n+1}^{-1}(1 - c + ce_{n+1}))z$$

$$= ca_{n+1}^{-1}(z - e_{n+1}z).$$

Since

$$a_{n+1}^{-1} = \phi(e_1)\phi(e_2)\cdots\phi(e_{n+1}),$$

$$||a_{n+1}^{-1}|| \le (1 - c - cK)^{-(n+1)},$$

and so

$$\sum_{n=0}^{\infty} ||y_{n+1} - y_n|| \le \sum_{n=0}^{\infty} cM(1 - c - cK)^{-(n+1)}||e_{n+1}z - z||.$$

We choose inductively e_{n+1} in A with $||e_{n+1}|| \leq K$ such that

$$||e_{n+1}a'_n - a'_n|| < \varepsilon_{n+1} \quad \text{and} \quad ||e_{n+1}z - z|| < \varepsilon_{n+1},$$

where

$$\varepsilon_{n+1} = \frac{1}{cM}(1 - c - cK)^{n+1} \frac{\delta}{2^{n+1}}.$$

Then

$$\sum_{n=0}^{\infty} ||y_{n+1} - y_n|| < \sum_{n=0}^{\infty} \frac{\delta}{2^{n+1}} = \delta.$$

It follows that $\{y_n\}$ is a Cauchy sequence in X and so has a limit $y \in \overline{Az}$. Since

$$y - z = \sum_{n=0}^{\infty} (y_{n+1} - y_n),$$

we see that

$$||y - z|| < \delta.$$

Also $\{a_n\}$ is a Cauchy sequence in A_1 since

$$\sum_{n=0}^{\infty} ||e_{n+1}a'_n - a'_n|| < \frac{1}{cM} \sum_{n=0}^{\infty} (1 - c - cK)^{n+1} \frac{\delta}{2^{n+1}} < \frac{\delta}{cM}.$$

If $a = \lim a_n$, then $a \in A$ since $a = \lim a'_n$. Furthermore, $z = \lim a_n y_n = ay$.

The inequality

$$||a_n|| \leq (1 - c + cK)^n, \quad n = 1,2,\ldots,$$

gives for $K = 1$ the precise estimate $||a|| \leq K = 1$. For arbitrary K we observe that

$$a = a_o + \sum_{n=0}^{\infty} (a_{n+1} - a_n) = \sum_{n=0}^{\infty} c(1-c)^n e_{n+1} + c \sum_{n=0}^{\infty} (e_n a_n' - a_n')$$

implies: $||a|| \leq K + \frac{\delta}{M} \leq K + \delta$. Thus we have a factorization $z = ay$ with $||a|| \leq K + \delta$ and $||y - z|| < \delta$, where δ is arbitrarily small. Let α be in $(0,1)$ and consider the factorization $z = (\alpha a)(\alpha^{-1}y)$. If δ is small and α is close to 1, then $||\alpha a|| \leq K$ and at the same time $\alpha^{-1}y$ is near z. Thus the theorem is proved. □

Finally, we give a reformulated version of Koosis' proof of the fundamental factorization theorem utilizing a generalized closed graph theorem.

Pták's generalized closed graph theorem. Let (E,d) be a complete metric space. Let $W(r)$, $0 < r < 1$, be a family of subsets of E such that there exist functions $q:(0,1) \to (0,\infty)$ and $\omega: R^+ \to (0,1)$ satisfying:

1) $\lim_{r \to 0^+} q(r) = 0$ and $\lim_{r \to 0^+} \omega(r) = 0$;

2) $\sigma(r) = \sum_{n=0}^{\infty} \omega^n(r) < \infty$ for all $r \in (0,1)$, (ω^n is the nth-iterate of ω);

3) $B(x,r) \cap W(q(\omega(r))) \neq \emptyset$ for all sufficiently small $r \in (0,1)$ and all $x \in W(q(r))$, where $B(x,r) = \{x' \in E: d(x',x) < r\}$.

Set

$$W = \bigcap_{0<s<1} \left(\bigcup_{r \leq s} W(r) \right)^{-}.$$

Then $B(x,\sigma(r)) \cap W \neq \emptyset$ for all $r \in (0,1)$ and all $x \in W(q(r))$.

Proof. Let $x \in W(q(r))$. Then by 3) we can choose inductively a sequence $\{x_n\}_{n=0}^{\infty}$ of elements in E such that: $x_o = x$,

$$x_{n+1} \in B(x_n,\omega^n(r)) \cap W(q(\omega^{(n+1)}(r))), \quad n = 0,1,2,\dots .$$

Since

$$\sum_{n=0}^{\infty} d(x_{n+1},x_n) < \sum_{n=0}^{\infty} \omega^n(r) = \sigma(r) < \infty,$$

it follows that $\{x_n\}_{n=0}^{\infty}$ is a Cauchy sequence in E. Thus $x_\infty = \lim x_n$ exists and $x_\infty \in B(x,\sigma(r)) \cap W$. □

Third proof. Fix a real number $c > 0$ such that

$$0 < K < \frac{1-c}{c};$$

for example take

$$c = \frac{1}{K+2} \quad \text{or} \quad c = \frac{1}{2K+1}.$$

For $e \in A$ with $||e|| \leq K$ we once again define

$$\phi(e) = (1 - c + ce)^{-1}.$$

Then $\phi(e)$ is a well defined element in A_1 with

$$||\phi(e)|| \leq (1 - c - cK)^{-1},$$

and

$$||\phi(e)x - x|| \leq M(1 - c - cK)^{-1}||ex - x|| \quad \text{for all } x \in X.$$

Fix any number δ, $0 < \delta < c$, and set $K_\delta = K(1 + \delta)$. Consider the complete metric space

$$E_\delta = \{a \in A: ||a|| \leq K_\delta\} \times \overline{Az},$$

where the distance in E_δ is determined by the norm

$$||(a,y)|| = \max\{||a||, \frac{1}{\delta}||y||\} \quad \text{for} \quad (a,y) \in E_\delta.$$

For r, $0 < r < 1$, define

$$W(r) = \{(a,y) \in E_\delta : ||a|| \leq K_\delta - Kr, \text{ and there exists a number } \alpha, \\ |\alpha| < r, \text{ such that } (a + \alpha)^{-1} \text{ exists in } A_1, \text{ and} \\ y = (a + \alpha)^{-1}z\}.$$

Now suppose $(a,y) \in W(r)$. Set

$$a' = (1 - c + ce)(a + \alpha) - (1 - c)\alpha;$$
$$y' = (a + \alpha)^{-1}(1 - c + ce)^{-1}z,$$

where the element $e \in A$ with $||e|| \leq K$ is chosen such that

$$||(e - 1)a|| + |\alpha|K < rK,$$

and

$$||(a + \alpha)^{-1}||M(1 - c - cK)^{-1}||(1 - e)z|| < \delta rK.$$

We claim that

$$(a',y') \in B((a,y),cKr) \cap W((1 - c)r).$$

Clearly $a' \in A$ and $y' \in \overline{Az}$. Since

$$a' + (1 - c)\alpha = (1 - c + ce)(a + \alpha),$$

the inverse $(a' + (1 - c)\alpha)^{-1}$ exists in A_1, $|(1 - c)\alpha| < (1 - c)r$, and

$$(a' + (1 - c)\alpha)y' = (1 - c + ce)(a + \alpha)(a + \alpha)^{-1}(1 - c + ce)^{-1}z = z.$$

Furthermore,

$$a' - a = (1 - c + ce)(a + \alpha) - (1 - c)\alpha - a$$
$$= c(e - 1)a + c\alpha e,$$

$$y' - y = (a + \alpha)^{-1}(1 - c + ce)^{-1}z - (a + \alpha)^{-1}z$$
$$= (a + \alpha)^{-1}(1 - c + ce)^{-1}c(1 - e)z,$$

and so

$$||a' - a|| \leq c||(e - 1)a|| + c|\alpha|K < crK,$$

$$\frac{1}{\delta}||y' - y|| \leq \frac{1}{\delta}||(a + \alpha)^{-1}||M(1 - c - cK)^{-1}c||(1 - e)z|| < crK;$$

finally,

$$||a'|| \leq ||a|| + ||a' - a|| \leq K_\delta - Kr + crK = K_\delta - (1 - c)Kr.$$

This proves the claim.

Setting

$$q(r) = \frac{r}{cK} \quad \text{and} \quad \omega(r) = (1 - c)r,$$

we see that the $W(r)$, $0 < r < 1$, satisfy the hypotheses of the generalized closed graph theorem. Observe that

$$\sigma(r) = \sum_{n=0}^{\infty} (1 - c)^n r = \frac{1}{c}r.$$

Setting $r_\delta = 1 - c + \delta$, we claim that

$$(ce, (1 - c + ce)^{-1}z) \in W(r_\delta)$$

for all $e \in A$ with $||e|| \leq K$. Indeed,

$$||ce|| \leq cK = K_\delta - Kr_\delta, \quad |1 - c| < r_\delta,$$

and

$$(1 - c + ce)^{-1} \text{ exists in } A_1.$$

Now choose $e_\delta \in A$ with $||e_\delta|| \leq K$ such that

$$||e_\delta z - z|| < \frac{1}{M}(1 - c - cK)\delta.$$

Then

$$(a_\delta, y_\delta) \equiv (ce_\delta, (1 - c + ce_\delta)^{-1}z) \in W(r_\delta) = W(q(cKr_\delta)),$$

and so, by the generalized closed graph theorem, there exists an element

$$(a_\infty, y_\infty) \in B((a_\delta, y_\delta), Kr_\delta) \cap W.$$

Thus $||a_\infty|| \leq K_\delta = K(1 + \delta)$, $y_\infty \in \overline{Az}$, $z = a_\infty y_\infty$, and

$$||y_\infty - z|| \leq ||y_\infty - y_\delta|| + ||y_\delta - z|| \leq \delta K r_\delta + ||\phi(e_\delta)z - z||$$

$$< \delta K + M(1 - c - cK)^{-1}||e_\delta z - z|| < \delta(K + 1).$$

Hence we have demonstrated the existence of a factorization $z = ay$ with $||a|| \leq K + \delta$, $y \in \overline{Az}$, and $||y - z|| < \delta$, and so the fundamental factorization theorem is proved. □

(16.2) <u>Corollary</u>. Let A be a Banach algebra with a left approximate identity bounded by $K \geq 1$. Then for every $z \in A$ and $\delta > 0$ there exist elements a and y in A such that:

i) $z = ay$;

ii) $||a|| \leq K$;

iii) $y \in \overline{Az}$;

iv) $||y - z|| < \delta$.

Thus, if a Banach algebra has either a bounded left or right approximate identity, then each of its elements can be factored. We will show later by example that the converse is false. Here we remark only that under suitable factorability hypotheses the converse is true.

(16.3) Proposition. Let A be a normed algebra with the following factorization property: there exists a constant $K \geq 1$ such that for every $z \in A$ and $\delta > 0$ there are elements $a, y \in A$ such that

$$z = ay, \quad ||a|| \leq K, \quad \text{and} \quad ||y - z|| < \delta.$$

Then A has a left approximate identity bounded by K.

Proof. For every $z \in A$ and every $\varepsilon > 0$ there are elements $a, y \in A$ such that

$$z = ay, \quad ||a|| \leq K, \quad \text{and} \quad ||y - z|| < \frac{\varepsilon}{K}.$$

Then

$$||z - az|| = ||ay - az|| \leq ||a||\cdot||y - z|| < \varepsilon.$$

Hence A has left approximate units bounded by K. By Theorem (9.3) A has a left approximate identity bounded by the same constant K. □

(16.4) Discussion. Let A be a Banach algebra, K some constant ≥ 1,

X a left Banach A-module, and $z \in X$. By Theorem (9.3) and Proposition (15.2) the following conditions are equivalent:

1. A has a left approximate identity bounded by K and $z \in X_e$;
2. For every finite set $\{a_1,\ldots,a_n\}$ of elements in A and every $\varepsilon > 0$ there is an element $e \in A$ with $||e|| \leq K$ such that

$$||ea_k - a_k|| < \varepsilon \quad (1 \leq k \leq n) \quad \text{and} \quad ||ez - z|| < \varepsilon.$$

3. For every $a \in A$ and every $\varepsilon > 0$ there is an element $e \in A$ with $||e|| \leq K$ such that

$$||ea - a|| < \varepsilon \quad \text{and} \quad ||ez - z|| < \varepsilon.$$

In our statement of the fundamental factorization theorem (16.1) for <u>left</u> Banach A-modules over a Banach algebra A with a bounded <u>left</u> approximate identity we assumed condition 1. The Cohen-Hewitt proof only made use of condition 2, and the variants of Koosis' proof used condition 3.

Now let us consider the factorization of <u>left</u> Banach A-modules over a Banach algebra A with a bounded <u>right</u> approximate identity. A slight modification of the Cohen-Hewitt proof will give such a factorization result under the hypothesis:

2'. For every finite set $\{a_1,\ldots,a_n\}$ of elements in A, every $x \in \overline{Az} \cup \{z\}$, and every $\varepsilon > 0$ there is an element $e \in A$ with $||e|| \leq K$ such that

$$||a_ke - a_k|| < \varepsilon \quad (1 \leq k \leq n) \quad \text{and} \quad ||ex - x|| < \varepsilon.$$

An analogous modification of the Koosis proof gives the same result under the formally weaker hypothesis:

3'. For every $a \in A$, every $x \in \overline{Az} \cup \{z\}$, and every $\varepsilon > 0$ there is an element $e \in A$ with $||e|| \leq K$ such that

$$||ae - a|| < \varepsilon \quad \text{and} \quad ||ex - x|| < \varepsilon.$$

Clearly, both conditions imply that A has a right approximate identity bounded by K and that $z \in X_e$. We do not know if the two conditions 2' and 3' are equivalent.

(16.5) Theorem. Let A be a Banach algebra and let X be a left Banach A-module. Suppose that the element $z \in X$ has the following property: there exists a constant $K \geq 1$ such that for every $a \in A$, every $x \in \overline{Az} \cup \{z\}$, and every $\varepsilon > 0$ there is an element $e \in A$ with $||e|| \leq K$ such that

$$||ae - a|| < \varepsilon \quad \text{and} \quad ||ex - x|| < \varepsilon.$$

Then for every $\delta > 0$ there exist elements $a \in A$ and $y \in X$ such that:

i) $z = ay$;

ii) $||a|| \leq K$;

iii) $y \in \overline{Az}$;

iv) $||y - z|| < \delta$.

First proof. (under the stronger assumption (16.4.2')). Fix a real number $c > 0$ such that

$$0 < K < \frac{1-c}{c};$$

for example take

$$c = \frac{1}{K+2} \quad \text{or} \quad c = \frac{1}{2K+1}.$$

For $e \in A$ with $||e|| \leq K$ define

$$\phi(e) = (1 - c + ce)^{-1}.$$

Then $\phi(e)$ is a well defined element in A_1 with

$$||\phi(e)|| \leq (1 - c - cK)^{-1},$$

and

$$||\phi(e)x - x|| \leq M(1 - c - cK)^{-1}||ex - x|| \quad \text{for all} \quad x \in X.$$

We shall choose inductively a sequence $\{e_n\}$ of elements in A with $||e_n|| \leq K$, $n = 1,2,\ldots$. Set

$$a_o = 1, \quad a_n = (1 - c)^n + \sum_{k=1}^{n} c(1 - c)^{k-1}e_k, \quad n = 1,2,\ldots .$$

Then

$$||a_{n+1}\phi(e_{n+1}) - a_n|| \leq (1 - c - cK)^{-1} \sum_{k=1}^{n} ||e_k e_{n+1} - e_k||.$$

If $(a_{n+1}\phi(e_{n+1}))^{-1}$ exists then a_{n+1}^{-1} exists. Setting

$$y_n = a_n^{-1}z, \quad n = 0,1,2,\ldots,$$

we have:

$$||y_{n+1} - y_n|| \leq ||\phi(e_{n+1})(a_{n+1}\phi(e_{n+1}))^{-1}z - \phi(e_{n+1})a_n^{-1}z||$$

$$+ ||\phi(e_{n+1})a_n^{-1}z - a_n^{-1}z||$$

$$\leq M||\phi(e_{n+1})||\cdot||(a_{n+1}\phi(e_{n+1}))^{-1} - a_n^{-1}||\cdot||z||$$

$$+ M(1 - c - cK)^{-1}||e_{n+1}a_n^{-1}z - a_n^{-1}z||.$$

Thus choosing inductively e_{n+1} in A with $||e_{n+1}|| \leq K$ such that

$$||e_k e_{n+1} - e_k|| \quad (1 \leq k \leq n) \quad \text{and} \quad ||e_{n+1}a_n^{-1}z - a_n^{-1}z||$$

are sufficiently small, we can conclude that a_{n+1}^{-1} exists in A_1 and

$$||y_{n+1} - y_n|| < \frac{\delta}{2^{n+1}} \quad \text{for all} \quad n = 0,1,2,\ldots .$$

Then $y = \lim y_n$ exists in X, $y \varepsilon \overline{Az}$, and $||y - z|| < \delta$. Furthermore,

$$a = \sum_{k=1}^{\infty} c(1 - c)^{k-1}e_k$$

is an element in A with $||a|| \leq K$ such that $z = ay$. □

<u>Second proof</u>. Fix a real number $c > 0$ such that

$$0 < K < \frac{1-c}{c};$$

for example take

$$c = \frac{1}{K+2} \quad \text{or} \quad c = \frac{1}{2K+1}.$$

For $e \varepsilon A$ with $||e|| \leq K$ define

$$\phi(e) = (1 - c + ce)^{-1}.$$

Then $\phi(e)$ is a well defined element in A_1 with

$$||\phi(e)|| \leq (1 - c - cK)^{-1}.$$

We shall choose inductively a sequence $\{e_n\}$ of elements in A with $||e_n|| \leq K$, $n = 1,2,\ldots$. Set

$$a_0 = 1, \quad a_{n+1} = a_n(1 - c + ce_{n+1}), \quad n = 0,1,2,\ldots .$$

Then

$$a_n = (1 - c)^n + a'_n, \quad \text{where} \quad a'_n \in A,$$

and

$$a_{n+1} - a_n = c(1 - c)^n(e_{n+1} - 1) + c(a'_n e_{n+1} - a'_n).$$

Now set

$$y_n = a_n^{-1} z, \quad n = 0,1,2,\ldots .$$

Then

$$\begin{aligned} y_{n+1} - y_n &= (a_{n+1}^{-1} - a_n^{-1})z \\ &= \phi(e_{n+1})(a_n^{-1} - (1 - c + ce_{n+1})a_n^{-1})z \\ &= \phi(e_{n+1})c(a_n^{-1}z - e_{n+1}a_n^{-1}z) \end{aligned}$$

and so

$$\sum_{n=0}^{\infty} ||y_{n+1} - y_n|| \leq \sum_{n=0}^{\infty} cM(1 - c - cK)^{-1}||a_n^{-1}z - e_{n+1}a_n^{-1}z||.$$

We choose inductively e_{n+1} in A with $||e_{n+1}|| \leq K$ such that

$$||a_n'e_{n+1} - a_n'|| < \varepsilon_{n+1} \quad \text{and} \quad ||a_n^{-1}z - e_{n+1}a_n^{-1}z|| < \varepsilon_{n+1},$$

where

$$\varepsilon_{n+1} = \frac{1}{cM}(1 - c - cK)\frac{\delta}{2^{n+1}}.$$

Then $y = \lim y_n$ exists in X, $y \in \overline{Az}$, and $||y - z|| < \delta$. Also $\{a_n\}$ is a Cauchy sequence in A_1 since

$$\sum_{n=0}^{\infty} ||a_n'e_{n+1} - a_n'|| < \frac{1}{cM}(1 - c - cK) \sum_{n=0}^{\infty} \frac{\delta}{2^{n+1}} < \frac{\delta}{cM}.$$

Thus $a = \lim a_n$ is an element in A such that $z = ay$.

The inequality

$$||a_n|| \leq (1 - c + cK)^n, \quad n = 1,2,\ldots,$$

gives for $K = 1$ the precise estimate

$$||a|| \leq K = 1.$$

For arbitrary K we observe that

$$a = a_o + \sum_{n=0}^{\infty} (a_{n+1} - a_n)$$

$$= \sum_{n=0}^{\infty} c(1 - c)^n e_{n+1} + c \sum_{n=0}^{\infty} (a_n'e_{n+1} - a_n')$$

implies

$$||a|| \leq K + \frac{\delta}{M} \leq K + \delta.$$

Thus we have shown the existence of a factorization $z = ay$

with

$$||a|| \leq K + \delta \quad \text{and} \quad ||y - z|| < \delta,$$

where δ can be an arbitrarily small real number. Let α be any number with $0 < \alpha < 1$ and consider the factorization

$$z = (\alpha a)(\alpha^{-1} y).$$

If δ is sufficiently small and α is sufficiently near to 1, then $||\alpha a|| \leq K$ and, at the same time, $\alpha^{-1}y$ is near by z. Thus the theorem is proved. □

A typical application of the fundamental factorization theorem for Banach modules is the following:

(16.6) <u>Example</u>. Let G be an arbitrary locally compact topological group, and let A be the group algebra $L^1(G)$. Then the Banach space $L^p(G)$, $1 \leq p < \infty$, is a left Banach A-module, where A acts on $L^p(G)$ by

$$L^1(G) \times L^p(G) \to L^p(G)$$

$$(f,g) \to f*g,$$

where $f*g$ denotes, as usual, the convolution of f and g. Properties i) - iii) of Definition (14.1) are routine verifications. The space $L^p(G)$ is, in fact, an <u>isometric</u> left Banach A-module, for it is well known and easily checked that

$$||f*g||_p \leq ||f||_1 ||g||_p, \quad f \in L^1(G),\ g \in L^p(G).$$

By the results in Section 13, A has a two-sided approximate identity bounded by one and $L^p(G)$ is essential. Thus by the fundamental factorization theorem

$$L^1(G)*L^p(G) = L^p(G).$$

The next example illustrates that the boundedness assumption for the approximate identity is essential in the fundamental factorization theorem.

(16.7) Example. Let p be a real number, $1 \leq p < \infty$. Let ℓ^p be the set of all sequences $a = (\alpha_1, \alpha_2, \ldots)$ of complex numbers for which the infinite series

$$\sum_{n=1}^{\infty} |\alpha_n|^p$$

converges. Then ℓ^p is a commutative Banach algebra under coordinatewise algebraic operations and norm defined by

$$||a||_p = \left(\sum_{n=1}^{\infty} |\alpha_n|^p\right)^{1/p}.$$

Clearly,

$$u_i = (1, \ldots, 1, 0, 0, \ldots), \quad i = 1, 2, \ldots,$$

is an unbounded approximate identity in ℓ^p. Thus $\ell^p \cdot \ell^p$ is a dense subset of ℓ^p. On the other hand,

$$(1,\frac{1}{2^{2/p}},\ldots,\frac{1}{n^{2/p}},\ldots)$$

is an element in ℓ^p which is not in $\ell^p.\ell^p$.

Thus ℓ^p, $1 \leq p < \infty$, is an example of a nonfactorable Banach algebra with an unbounded approximate identity.

17. Multiple factorization.

(17.1) Theorem. Let A be a Banach algebra with a left approximate identity $\{e_\lambda\}$ bounded by $K \geq 1$ and let X be a left Banach A-module. Suppose Z is a bounded subset of X such that

$$||e_\lambda z - z|| \to 0 \quad \text{uniformly for} \quad z \in Z.$$

Then for every $\delta > 0$ there exists an element $a \in A$ and a subset Y of X such that:

i) $Z = aY$;

ii) $||a|| \leq K$;

iii) $Y \subset \overline{AZ}$;

iv) $||y - ay|| < \delta \quad (y \in Y)$;

v) Y is bounded;

vi) The map $y \to ay$ is a homeomorphism of Y onto Z.

Proof. We refer to the Cohen-Hewitt proof of the fundamental factorization theorem (16.1). Choose inductively e_{n+1} in A with $||e_{n+1}|| \leq K$ and

$$||e_{n+1}e_k - e_k|| \quad (1 \leq k \leq n) \quad \text{and} \quad ||e_{n+1}z - z|| \quad (z \in Z)$$

sufficiently small such that a_{n+1}^{-1} exists in A_1 and

$$||a_{n+1}^{-1}z - a_n^{-1}z|| < \frac{\delta}{2^{n+1}} \quad \text{for all} \quad z \in Z, \quad n = 0,1,2,\ldots .$$

Then

$$a = \sum_{k=1}^{\infty} c(1-c)^{k-1}e_k \quad \text{and} \quad Y = \{y = \lim a_n^{-1}z\colon z \in Z\}$$

satisfy i) - v). Since $a_n^{-1}z \to y$ uniformly on Z, we see that the map $y \to ay$ is a homeomorphism of Y onto Z. □

Here is an important alternate proof of (17.1) which applies the fundamental factorization theorem to an auxiliary left Banach A-module. Its simplicity shows that the preceding proof is of historical value only.

Let V be the set of all continuous functions from Z into X_e whose range is a bounded subset of X_e on which $\{e_\lambda\}$ converges uniformly to the identity. It is routine to verify that V is a left Banach A-module under pointwise algebraic operations and sup-norm:

$$||f||_\infty = \sup\{||f(z)||\colon z \in Z\}$$

Obviously,

$$||e_\lambda f - f||_\infty \to 0 \quad \text{for each} \quad f \in V.$$

We see that the fundamental factorization theorem (16.1) applies, and defining $f \in V$ by $f(z) = z$ for $z \in Z$, there exists $a \in A$ and $g \in V$ so that

$$f = ag, \; ||a|| \leq K, \; g \in \overline{Af}, \text{ and } ||g - ag||_\infty < \delta.$$

It is now easy to check that a and

$$Y = \{g(z): z \in Z\}$$

satisfy i) - vi). □

Recall that a subset Z of a normed linear space X is said to be <u>totally bounded</u> if for every $\varepsilon > 0$ there exist finitely many elements $z_1, z_2, \ldots, z_n$ in X such that given any $z \in Z$ we can find a z_k such that $||z - z_k|| < \varepsilon$. Clearly, every compact subset is totally bounded, and every totally bounded subset is uniformly bounded.

(17.2) <u>Lemma</u>. Let A be a Banach algebra with a left approximate identity $\{e_\lambda\}_{\lambda \in \Lambda}$ bounded by $K \geq 1$ and let X be a left Banach A-module. If Z is a totally bounded subset of X_e then

$$||e_\lambda z - z|| \to 0 \text{ uniformly for } z \in Z.$$

<u>Proof</u>. Let $\varepsilon > 0$. Then there exist finitely many elements $z_1, z_2, \ldots, z_n$ in X_e such that given any $z \in Z$ we can find a z_k such that $||z - z_k|| < \varepsilon$. Now choose $\lambda_o \in \Lambda$ such that

$$||e_\lambda z_k - z_k|| < \varepsilon \quad (1 \leq k \leq n) \quad \text{for all } \lambda \geq \lambda_o.$$

Then given any $z \in Z$,

$$||e_\lambda z - z|| \leq M||e_\lambda|| \cdot ||z - z_k|| + ||e_\lambda z_k - z_k|| + ||z_k - z||,$$

and so

$$||e_\lambda z - z|| < (2 + MK)\varepsilon \quad \text{for all} \quad \lambda \geq \lambda_o.$$

Thus

$$||e_\lambda z - z|| \to 0 \quad \text{uniformly for} \quad z \in Z. \square$$

(17.3) Theorem. Let A be a Banach algebra with a left approximate identity bounded by $K \geq 1$ and let X be a left Banach A-module. If Z is a compact subset of X_e, then for every $\delta > 0$ there exists an element $a \in A$ and a subset Y of X such that:

i) $Z = aY$;

ii) $||a|| \leq K$;

iii) $Y \subset \overline{AZ}$;

iv) $||y - ay|| < \delta \quad (y \in Y)$;

v) Y is compact;

vi) The map $y \to ay$ is a homeomorphism of Y onto Z.

Proof. By Lemma (17.2) we may apply Theorem (17.1). Since Y is homeomorphic to Z, also Y is compact.

Alternatively, consider the left Banach A-module of all continuous functions from Z into X_e and apply the fundamental factorization theorem (16.1). $\square$

(17.4) Theorem. Let A be a Banach algebra with a left approximate identity bounded by $K \geq 1$ and let X be a left Banach A-module. If $\{z_n\}_{n=1}^{\infty}$ is a convergent sequence of elements in X_e, then for every $\delta > 0$ there exists an element $a \in A$ and a sequence $\{y_n\}_{n=1}^{\infty}$ of elements in X such that for all $n = 1,2,...,$

i) $z_n = ay_n$;

ii) $||a|| \leq K$;

iii) $y_n \in \overline{Az_n}$;

iv) $||y_n - z_n|| < \delta$;

v) $\{y_n\}_{n=1}^{\infty}$ is convergent.

Proof. Set $z_\infty = \lim z_n$ and apply Theorem (17.3) to the compact subset $Z = \{z_n : n = 1,2,\ldots,\infty\}$. Since $Y = \{y_n : n = 1,2,\ldots,\infty\}$ is homeomorphic to Z, the sequence $\{y_n\}_{n=1}^{\infty}$ is also convergent. The fact that $y_n \in \overline{Az_n}$ for each $n = 1,2,3,\ldots$ follows from the proofs of (17.1) and (17.3).

Alternatively, consider the left Banach A-module of all functions f from the set of natural numbers into X_e such that $\{f(n)\}_{n=1}^{\infty}$ is convergent and apply the fundamental factorization theorem (16.1). □

(17.5) Theorem. Let A be a Banach algebra with a left approximate identity bounded by $K \geq 1$ and let X be a left Banach A-module. If $\{z_n\}_{n=1}^{\infty}$ is a sequence of elements in X_e with $\lim z_n = 0$, then for every $\delta > 0$ there exists an element $a \in A$ and a sequence $\{y_n\}_{n=1}^{\infty}$ of elements in X such that for all $n = 1,2,\ldots,$

i) $z_n = ay_n$;

ii) $||a|| \leq K$;

iii) $y_n \in \overline{Az_n}$;

iv) $||y_n - z_n|| < \delta$;

v) $\lim y_n = 0$.

Proof. Set $z_\infty = 0$ and apply Theorem (17.3) to the compact subset

$Z = \{z_n : n = 1,2,\ldots,\infty\}$. Since the set $Y = \{y_n : n = 1,2,3,\ldots,\infty\}$ is homeomorphic to Z under the homeomorphism $y \to ay$ (see the proof of theorem (17.1)), and since z_n is sent to y_n, then $z_\infty = 0$ implies that $y_\infty = \lim y_n = 0$.

Alternatively, consider the left Banach A-module of all functions f from the set of natural numbers into X_e such that $\lim f(n) = 0$ and apply the fundamental factorization theorem (16.1)· □

(17.6) Corollary. Let A be a Banach algebra with a left approximate identity bounded by $K \geq 1$ and let X be a left Banach A-module. If Z is a countable subset of X_e, then there exists an element $a \in A$ and a countable subset Y of X such that

$$Z = aY, \ ||a|| \leq K, \text{ and } Y \subset \overline{AZ}.$$

Proof. Write Z as a sequence $\{x_n\}_{n=1}^{\infty}$ and apply Theorem (17.5) to the sequence $\{z_n\}_{n=1}^{\infty}$, where

$$z_n = \frac{1}{1+n||x_n||} x_n. \quad \square$$

Remark. As important applications of these multiple factorization results we shall prove Varopoulos' theorem on the continuity of positive functionals, B. E. Johnson's theorem on continuity of centralizers, and, finally, a recent theorem of M. D. Green which shows that every maximal left (or right) ideal is closed in a Banach algebra with bounded approximate identity.

Recall that a positive functional on a *-algebra A is a linear functional p on A such that $p(x^*x) \geq 0$ for all $x \in A$.

(17.7) Theorem. (N. Th. Varopoulos). Let A be a Banach *-algebra with a bounded approximate identity. Then every positive functional p on A is continuous.

Proof. For a fixed element $a \in A$ consider the positive functional ${}_{a^*}p_a$ on A defined by

$${}_{a^*}p_a(x) = p(a^*xa),\ x \in A.$$

Since ${}_{a^*}p_a$ can be extended to a positive functional on A_1 (simply set ${}_{a^*}p_a(x + \alpha) = p(a^*(x + \alpha)a),\ x + \alpha \in A_1$), ${}_{a^*}p_a$ is continuous. For all $a, x, b \in A$ the polar identity

$$4axb = \sum_{k=0}^{3} i^k(a^* - ib)^*x(a^* - ib)$$

holds. Therefore, the linear functional ${}_ap_b$ on A defined by

$${}_ap_b(x) = p(axb),\ x \in A,$$

is also continuous.

Now let $\{z_n\}$ be a sequence in A such that $\lim z_n = 0$. Applying the left and right versions of (17.5) to the essential Banach A-module A we find elements $a, b \in A$ and sequences $\{y_n\}$ and $\{w_n\}$ in A with limit 0 such that $z_n = ay_n = aw_nb$ for all n. Then the continuity of ${}_ap_b$ gives $\lim p(z_n) = \lim {}_ap_b(w_n) = 0$, and so p is continuous. □

(17.8) Theorem. (B. E. Johnson). Let A be a Banach algebra with a bounded left approximate identity and let X be a left Banach A-module. If T is a function from A into X such that $T(ab) = aT(b)$ for all $a, b \in A$, then T is a bounded linear transformation.

Proof. Let a_1, $a_2 \in A$ and $\lambda \in C$. Apply the right module version of (17.5), with $X = A$, to obtain a, b_1, $b_2 \in A$ with $a_1 = b_1 a$ and $a_2 = b_2 a$. Then

$$T(a_1 + a_2) = T((b_1 + b_2)a) = (b_1 + b_2)T(a)$$
$$= b_1 T(a) + b_2 T(a) = T(b_1 a) + T(b_2 a) = T(a_1) + T(a_2).$$

Thus T is linear. To see that T is continuous, assume $\{a_n\}$ is a sequence in A such that $a_n \to 0$. By the right version of (17.5), there is an $a \in A$ and a sequence $\{b_n\}$ in A satidfying $a_n = b_n a$ and $b_n \to 0$. Then $T(a_n) = b_n T(a) \to 0$. □

Lemma. Let A be an algebra, I a maximal left ideal of A, and $b \in A$. Then $J = \{a \in A: ab \in I\}$ is a maximal modular left ideal or it is equal to A.

Proof. Clearly, J is a left ideal. If $J \neq A$, there is $a_o \in A$ such that $a_0 b \notin I$ and hence $Ab + I = A$, since I is maximal. Then $eb + i = b$ for some $e \in A$, $i \in I$, and so $(a - ae)b = a(b - eb) = ai \in I$ for $a \in A$. Hence $a - ae \in J$ for all $a \in A$, so e is a modular right unit for J.

To see that J is maximal, let K be a left ideal of A properly containing J. Then $Kb \not\subset I$ (otherwise $K \subset J$ by definition of J) so $Kb + I = A$ since I is maximal. As above, there is a modular right unit, say u, for J in K, i.e., $a - au \in J \subset K$ for $a \in A$. But $au \in aK \subset K$, so $a \in K$ for all $a \in A$. □

(17.9) Theorem. (M. D. Green). Let A be a Banach algebra with a bounded right approximate identity. Then every maximal left ideal I in A is closed.

Proof. Suppose $a_n \in I$ and that $a_n \to a$. We must show that $a \in I$. Since $a_n \to a$, the right version of (17.4) implies there are y_n, $b \in A$ with $y_n \to y$ for some $y \in A$ and $a_n = y_n b$ for each n. Then $a = \lim a_n = \lim y_n b = yb$. Now, by the lemma, $J = \{a \in A: ab \in I\}$ is a maximal modular left ideal or coincides with A. In either case J is closed. Since $a_n \in I$, then $y_n b \in I$ and so $y_n \in J$. Hence, $y \in J$, i.e., $a = yb \in I$. □

(17.10) Corollary. If A is a Banach algebra with a bounded two-sided approximate identity, then every maximal left (right) ideal in A is closed.

It was conjectured by C. A. Akemann and M. Rosenfeld [6] that if A is a Banach algebra which factors, then every maximal left (right) ideal in A is closed. The next example shows that this conjecture is false.

(17.11) Example. Let B be an infinite-dimensional Banach space. Let $A = B \oplus C$, with norm $||(x,\lambda)|| = ||x|| + |\lambda|$ and multiplication

$$(x,\lambda)(y,\mu) = (\lambda y,\lambda\mu) \qquad (x,\ y \in B;\ \lambda,\ \mu \in C).$$

Then A is a Banach algebra with left identity $(0,1)$ and hence A factors. From the definition of multiplication, it follows that every linear subspace of A is a left ideal in A. Let J be any maximal subspace of A which is not closed. Then J is a maximal left ideal in A which is not closed.

18. Power factorization.

In this section we begin to study some of the more intricate refinements and extensions of the fundamental factorization theorem (16.1). We show that an element z in a Banach algebra A with bounded left approximent identity may be factored as $z = a^n y_n$ for elements a, y_n in A with some control on the growth of the sequence $\{||y_n||\}$ of norms ($n = 1,2,3,\ldots$). More precisely:

§18. POWER FACTORIZATION

(18.1) Theorem. Let A be a Banach algebra with left approximate identity bounded by $K \geq 1$ and let X be a left Banach A-module. Let $\{\alpha_n\}$ be a sequence of real numbers such that $\alpha_n > 1$ for all n and $\alpha_n \to \infty$ as $n \to \infty$, let $\delta > 0$, and let N be a positive integer. Then for every $z \in X_e$ $(= \overline{AX})$, there exist elements $a \in A$ and $y_1, y_2, y_3, \ldots$ in X such that:

i) $z = a^j y_j$ for $j = 1,2,\ldots$;

ii) $||a|| \leq K$;

iii) $y_j \in (Az)^-$ for $j = 1,2,\ldots$;

iv) $||z - y_h|| < \delta$ for $h = 1,\ldots,N$, and

v) $||y_j|| \leq \alpha_j^j ||z||$ for $j = 1,2,\ldots$.

In the following two lemmas A is a Banach algebra with left approximate identity bounded by K, Y is a left Banach A-module, and Y_e is the essential part of Y. As usual we can regard Y as a Banach A_1-module (see (14.3)).

(18.2) Lemma. If $z_1,\ldots,z_n \in Y_e$ and $\varepsilon > 0$, then there exists an element $e \in A$ such that $||e|| \leq K$ and $||z_i - ez_i|| < \varepsilon$ $(1 \leq i \leq n)$.

Proof. See (15.2) and (1.2). □

(18.3) Lemma. (a) Let $0 < \lambda < (K + 1)^{-1}$. If $e \in A$ with $||e|| \leq K$, then $(1 - \lambda + \lambda e)^{-1}$ exists in A_1 and $||(1 - \lambda + \lambda e)^{-1}|| \leq (1 - \lambda - K\lambda)^{-1}$.

(b) If $\varepsilon > 0$, there is an $\eta > 0$ such that if $e \in A$, $||e|| \leq K$, $y \in Y$, $||y - ey|| < \eta$, then $||y - (1 - \lambda + \lambda e)^{-1}y|| < \varepsilon$.

Proof. Since $1 - \lambda$ is positive and $||\lambda(1 - \lambda)^{-1}e|| < \lambda(1 - \lambda)^{-1}K < 1$, the element $(1 + \lambda(1 - \lambda)^{-1}e)$ is invertible in A_1. Standard estimates using geometric series imply that

$$||(1 - \lambda + \lambda e)^{-1}|| = (1 - \lambda)^{-1}||(1 + \lambda(1 - \lambda)^{-1}e)^{-1}|| \le (1 - \lambda - \lambda K)^{-1}.$$

Hence

$$\begin{aligned} ||y - (1 - \lambda + \lambda e)^{-1}y|| &\le M(1 - \lambda - \lambda K)^{-1}||(1 - \lambda + \lambda e)y - y|| \\ &\le M\lambda(1 - \lambda - \lambda K)^{-1}||y - ey||. \end{aligned}$$

Taking $\eta = \varepsilon M^{-1}\lambda^{-1}(1 - \lambda - \lambda K)$ completes the proof. □

The idea behind the proof of (18.1) again goes back to Cohen. We shall apply Lemma (18.3) inductively to construct a sequence $\{a_n\}$ in A_1 that converges to an element $a \in A$, and such that $\{a_n^{-j}z\}$ is Cauchy in X for each fixed j even though the sequence $\{||a_n^{-j}||\}$ is unbounded for each fixed j. The control on the growth of the sequence $\{y_n\}$ is obtained by considering a subsequence $\{\alpha_{H(n)}\}$ of $\{\alpha_n\}$ that diverges to infinity fast, and doing the construction for the j-th powers on the intervals $[H(n),H(n+1)]$.

Proof of Theorem (18.1): We shall assume that $||z|| = 1$, and that

$$\delta \le \min \{1, \alpha_n^n - 1: n = 1,2,...\}.$$

We choose and fix a λ satisfying $0 < \lambda < (K + 1)^{-1}$. Next choose $H(0)$ so that $H(0) \ge N$ and for all $j \ge H(0)$ the inequality

$$\alpha_j > 2M(1 - \lambda - \lambda K)^{-1} + 1$$

holds. The sequence $\{H(n)\}$ of positive integers is now chosen so that $H(n)$ is the maximum of $H(n-1) + 1$ and

$$\inf \{j: \alpha_h \geq M \cdot 2^n(1 - \lambda - \lambda K)^{-n} + 1 \text{ for all } h \geq j\}.$$

The choice of the sequence $\{H(n)\}$ satisfying these conditions is possible because $\alpha_n > 1$ for all n and $\alpha_n \to \infty$ as $n \to \infty$.

We shall inductively define a sequence $\{e_n\}$ in A and a sequence $\{a_n\}$ of invertible elements in A_1 such that $a_o = 1$, $||e_n|| \leq K$,

$$a_n = \sum_{k=1}^{n} \lambda(1 - \lambda)^{k-1} e_k + (1 - \lambda)^n, \tag{1}$$

and

$$||a_n^{-j}z - a_{n-1}^{-j}z|| < \frac{\delta}{2^n} \tag{2}$$

for all $j \leq H(n)$ and all positive integers n. We may choose e_1 and a_1 to satisfy (1) and (2) by applying Lemmas (18.2) and (18.3) to X.

Suppose $e_o, \ldots, e_n$ have been chosen. Let $B = \{e \in A: ||e|| \leq K\}$ and define $F: B \to A_1$ by

$$F(e) = \sum_{k=1}^{n} \lambda(1 - \lambda)^{k-1}(1 - \lambda + \lambda e)^{-1} e_k + (1 - \lambda)^n.$$

Then

$$F(e) - a_n = \sum_{k=1}^{n} \lambda(1 - \lambda)^{k-1}((1 - \lambda + \lambda e)^{-1} e_k - e_k).$$

By Lemma (18.3) applied with Y equal to the direct sum of n copies of A

regarded as a left Banach A-module in the natural manner, $||F(e) - a_n||$ may be made arbitrarily small provided that $||ee_k - e_k||$ is sufficiently small for $k = 1,\dots,n$. Since $Inv(A_1)$ is open and the mapping $g \to g^{-1}$ is continuous on $Inv(A_1)$, it follows that $F(e)$ is invertible and $||F(e)^{-1} - a_n^{-1}||$ is arbitrarily small provided that $||ee_k - e_k||$ is sufficiently small for $k = 1,\dots,n$. We now apply Lemma (18.2), and Lemma (18.3) with Y equal to the direct sum of n copies of A and $H(n+1)$ copies of Y_e. We choose e_{n+1} with $||e_{n+1}|| \leq K$ so that $||e_k - e_{n+1}e_k||$ $(k = 1,\dots,n)$ and $||a_n^{-j}z - e_{n+1}a_n^{-j}z||$ $(j = 1,\dots,H(n))$ are so small that $F(e_{n+1})$ is invertible in A_1, $||F(e_{n+1})^{-1}|| \leq ||a_n^{-1}|| + 1$,

$$||F(e_{n+1})^{-1} - a_n^{-1}|| \leq \eta M^{-1}(2^{-n-1}(1 - \lambda - \lambda K)^{n+1})^{H(n)} \tag{3}$$

and

$$||\{(1 - \lambda + \lambda e_{n+1})^{-1} - 1\}a_n^{-j}z|| \leq \eta M^{-1}2^{-1}(||a_n^{-1}|| + 1)^{-1} \tag{4}$$

for $j = 1,\dots,H(n+1)$ where η, to be chosen later, does not depend on e_{n+1} or a_{n+1}. Then

$$\begin{aligned} a_{n+1} &= (1 - \lambda)^{n+1} + \lambda(1 - \lambda)^n e_{n+1} + \sum_{k=1}^{n} \lambda(1 - \lambda)^{k-1}e_k \\ &= (1 - \lambda + \lambda e_{n+1})F(e_{n+1}) \end{aligned}$$

and so

$$\begin{aligned} ||a_{n+1}^{-1}|| &\leq ||(1 - \lambda + \lambda e_{n+1})^{-1}||\cdot||F(e_{n+1})^{-1}|| \\ &\leq (1 - \lambda - \lambda K)^{-1}(||a_n^{-1}|| + 1). \end{aligned}$$

Because of the term $(1-\lambda)^{-n}$ in a_n^{-1} we have $||a_n^{-1}|| \geq (1-\lambda)^{-n} \geq 1$ which gives

$$||a_{n+1}^{-1}|| \leq 2(1-\lambda-\lambda K)^{-1}||a_n^{-1}||.$$

Repeated use of this formula and $a_o = 1$ leads to

$$||a_{n+1}^{-1}|| \leq 2^{n+1}(1-\lambda-\lambda K)^{-n-1}.$$

Let $1 \leq r \leq H(n)$. Then

$$\begin{aligned} ||(a_{n+1}^{-1} - a_n^{-1})a_n^{-r}z|| &= ||\{F(e_{n+1})^{-1}(1-\lambda+\lambda e_{n+1})^{-1} - a_n^{-1}\}a_n^{-r}z|| \\ &\leq M||F(e_{n+1})^{-1}||\cdot||\{(1-\lambda+\lambda e_{n+1})^{-1} - 1\}a_n^{-r}z|| + \\ &\quad + M||F(e_{n+1})^{-1} - a_n^{-1}||\cdot||a_n^{-r}|| \leq \eta \end{aligned}$$

by (3), (4), and the bound for $||a_n^{-1}||$. Hence

$$\begin{aligned} ||a_{n+1}^{-j}z - a_n^{-j}z|| &\leq \sum_{r=0}^{j-1} M||a_{n+1}^{-j+1+r}||\cdot||(a_{n+1}^{-1} - a_n^{-1})a_n^{-r}z|| \\ &\leq \sum_{r=0}^{j-1} M\cdot\{2^{n+1}(1-\lambda-\lambda K)^{-n-1}\}^{H(n)}\cdot\eta \\ &= j\cdot M\cdot(2^{n+1}(1-\lambda-\lambda K)^{-n-1})^{H(n)}\cdot\eta \\ &\leq \delta\cdot 2^{-n-1}, \end{aligned}$$

provided η is small enough, for $j = 1,\ldots,H(n)$. This completes the inductive construction of the sequences $\{e_n\}$ and $\{a_n\}$ satisfying (1) and (2).

The sequence $\{a_n\}$ is Cauchy in A_1, and $\lim_{n\to\infty} a_n = a$ is in A because $\lim_{n\to\infty} a_n = \sum_{k=1}^{\infty} \lambda(1-\lambda)^{k-1} a_k$. Further $||a|| \leq K$. The sequence $\{a_n^{-j}z\}$ is Cauchy in n for each j, since $n > m$ and $H(m+1) \geq j$ imply that

$$||a_n^{-j}z - a_m^{-j}z|| \leq \sum_{k=m}^{n-1} ||a_{k+1}^{-j}z - a_k^{-j}z|| \leq \delta/2^m \tag{5}$$

by (2). We let $y_j = \lim_{n\to\infty} a_n^{-j}z$ for each j. Then $z = a^j y_j$ for each j. Since z is in $(Az)^-$, by Theorem (16.1) we have $(A_1 z)^- = (Az)^-$ so that $y_j \in (Az)^-$. By (5) we have $||y_j - a_m^{-j}z|| \leq \delta/2^m$ if $H(m+1) \geq j$. Hence $||y_j - z|| \leq \delta$ for $j = 1,\ldots,N$ because $H(0) \geq N$ and $a_o = 1$.

If $1 \leq j \leq H(0)$, then $||y_j|| \leq 1 + \delta$ so that $||y_j|| \leq \alpha_j^j$ by the restriction on δ. Now suppose that $H(m) \leq j < H(m+1)$. Then the choice of $H(m)$ implies that $\alpha_j \geq M\cdot 2^m(1-\lambda-\lambda K)^{-m} + 1$. Also

$$||y_j|| \leq ||a_m^{-j}z|| + \delta \leq M||a_m^{-1}||^j + 1$$

$$\leq M\cdot 2^{mj}(1-\lambda-\lambda K)^{-jm} + 1$$

by the bound on $||a_m^{-1}||$. Hence

$$||y_j|| \leq \{M\cdot 2^m(1-\lambda-\lambda K)^{-m} + 1\}^j \leq \alpha_j^j$$

by the choice of $H(m+1) > j$. This completes the proof of Theorem (18.1).□

§19. FACTORIZATION AND SEMIGROUPS

19. Factorization and semigroups.

Let A be a Banach algebra with bounded approximate identity. We know from (16.1) that if $z \in A$ then there is an element $a \in A$ such that $z \in aA$. Further we have seen in Section 18 that a generalization is obtained if we replace $z \in aA$ by $z \in \bigcap_{n=1}^{\infty} a^n A$. In 1975 B. E. Johnson obtained an (unpublished) factorization $z = ay$ such that for each positive integer n there is an element $a^{1/n}$ in the algebra with $(a^{1/n})^n = a$. The main result in this section is a result of this type. If z is an element of a Banach algebra A with bounded approximate identity, then there are functions $t \to a^t$ and $t \to z_t$ from the set of positive real numbers R^+ into A such that $z = a^t z_t$ for all positive t, and $t \to a^t$ is a norm continuous one parameter semigroup.

It is clear that a Banach algebra A has a bounded two-sided approximate identity bounded by K if and only if for each $\varepsilon > 0$ and each finite subset $\{a_1,\ldots,a_n\}$ of A there is an $e \in A$ with $||e|| \leq K$ and $||ea_j - a_j|| + ||a_je - a_j|| < \varepsilon$ for $j = 1,\ldots,n$.

(19.1) Theorem. Let A be a Banach algebra with bounded two-sided approximate identity bounded by K, let X be a left Banach A-module, and let Y be a right Banach A-module. Let H denote the open right half $\{\lambda \in C\colon \operatorname{Re}\lambda > 0\}$ of the complex plane and let B be a bounded subset of H, let $\delta > 0$, and let $t \to \alpha_t$ be a continuous function from R^+ into $[1+\delta, \infty)$ such that $\alpha_t \to \infty$ as $t \to \infty$. If $z \in X_e$ $(= \overline{AX})$ and $y \in Y_e$ $(= \overline{YA})$, then there are analytic functions $t \to a^t$, $t \to z_t$, and $t \to y_t$ from H into A, X, and Y, respectively, such that:

i) $z = a^t z_t$ and $y = y_t a^t$ for all $t \in H$;

ii) $t \to a^t$ is a homomorphism from the additive semigroup H into the multiplicative semigroup A;

iii) $a^t z_{s+t} = z_s$ and $y_{s+t} a^t = y_s$ for all $s, t \in H$;

iv) If $K = 1$, then $||a^t|| \leq 1$ for all $t \in R^+$;

v) $a^t z \to z$ and $ya^t \to y$ as $t \to 0$ $(t \in H)$;

vi) $\{||a^t||: 0 < t \leq 1\}$ is bounded;

vii) $||z - z_t|| \leq \delta$ and $||y - y_t|| \leq \delta$ for all $t \in B$;

viii) $||z_t|| \leq \alpha^{|t|}_{|t|}$ and $||y_t|| \leq \alpha^{|t|}_{|t|}$ for all $t \in H$, and

xi) $z_t \in (Az)^-$ and $y_t \in (yA)^-$ for all $t \in H$.

The proof of (19.1), to be given below, is based on the first (Cohen-Hewitt) proof of (16.1) with variations to help us with powers. We wish to indicate here where the proof differs from the first proof of (16.1). Let A_1 be the Banach algebra obtained from A by adjoining an identity. In the Cohen-Hewitt proof (and also in (18.1)) a sequence $a_n = \sum_{j=1}^{n} \lambda(1 - \lambda)^{j-1} e_j + (1 - \lambda)^n$ is defined in A_1 such that a_n is invertible and the sequence $\{a_n^{-1} z\}$ is Cauchy in X. The sequence $\{a_n\}$ converges to an element in A by the choice of λ in relation to K. In the definition of the sequence $\{a_n\}$ a crucial role is played by the factors of the form $(1 + \lambda(e - 1))$. The difficulty when considering powers in the noncommutative case is that $(xy)^2 \neq x^2 y^2$. This difficulty is overcome by using the exponential function $\exp x = \sum_{n=0}^{\infty} x^n/n!$, and heuristically replacing the factor $(1 + \lambda(e - 1))$ by $\exp (e - 1)$. We use $a_n = \exp(\sum_{j=1}^{n} (e_j - 1))$. It is easy to obtain a^t from a_n^t, where (in general) we define d^t by $d^t = \exp tf$ for $t \in C$ when d has been defined by $d = \exp f$. In this proof the a_n are clearly invertible, and are chosen to be convergent whereas in (16.1) the a_n are clearly convergent and are chosen to be invertible. In the previous proofs of the factorization theorems, and in this one, the elements a_n are in the principal component of the group of invertibles of A_1. This is further motivation for the use of the exponential function. The elements a_n could be replaced by $a_n = \exp(\sum_{j=1}^{n} \lambda_j(e_j - 1))$, where $\{\lambda_j\}$ is a sequence of positive real numbers satisfying certain conditions but this does not seem to give further information. The reader will note that the terms $(e - 1)$ play a crucial role in all proofs of (16.1) because for each b there is an e with $||(e - 1)b||$ small.

We shall often need to choose an element e in A such that $||ea_j - a_j||$ and $||ey_j - y_j||$ are small for a finite number of j, where $a_j \in A$ and $y_j \in X_e$ by using a lemma that enables us to choose e for one element in a left Banach A-module. We shall make this choice by applying the lemma to the left Banach A-module $A \oplus \ldots \oplus A \oplus X \oplus \ldots \oplus X$, and the element $(a_1,\ldots,a_n,x_1,\ldots,x_n)$. In the proofs in this section we shall assume that our left and right Banach A-modules have been normalized to satisfy $||az|| \leq ||a||\cdot||z||$ and $||wa|| \leq ||a||\cdot||w||$ for $a \in A$ and z and w in the modules.

<u>Preliminary lemmas and remarks to the proof of Theorem</u> (19.1).

In the following three lemmas A is a Banach algebra with a bounded left approximate identity bounded by $K \geq 1$, X is a left Banach A-module, and X_e is the essential part of X. We shall also apply the lemmas to right Banach A-modules. The "right" version of the lemmas can be proved in a similar manner or obtained by considering the reversed product on the algebra and module.

(19.2) Lemma. Let n be a positive integer and let $\varepsilon > 0$.

(a) If $f \in A$, there is an $\eta > 0$ such that, if $e \in A$ with $||e|| \leq K$, $||f(e-1)|| + ||(e-1)f|| < \eta$, and $1 \leq k \leq n$, then $||f^k + (e-1)^k - (f + (e-1))^k|| < \varepsilon$.

(b) If $c = f + \mu 1 \in A_1$ ($f \in A$, $\mu \in C$) and $z \in X_e$, there is an $\eta > 0$ such that if $e \in A$ with $||e|| \leq K$, $||(e-1)z|| + ||(e-1)f|| < \eta$, and $1 \leq k \leq n$, then $||c^k z - (c + (e-1))^k z|| < \varepsilon$.

Proof. (a) Multiplying out the power $(f + (e-1))^k$ and using the norm inequalities we obtain

$$||f^k + (e-1)^k - (f + (e-1))^k|| \leq$$

$$\sum_{j=1}^{k-1} ||f||^{k-1-j} ||e-1||^{j-1} \{\binom{k}{j} - 1) ||(e-1)f|| + ||f(e-1)||\} \quad (1 \leq k \leq n).$$

We now choose $\eta > 0$ such that $\sum_{j=1}^{k-1} \binom{k}{j} ||f||^{k-1-j} (K+1)^{j-1} < \eta^{-1}\varepsilon$ for $k = 1,\ldots,n$. This proves (a).

(b) Multiplying out the powers $(\mu 1 + f)^k$ and $(\mu 1 + (f + (e-1)))^k$ and using the norm inequalities, we obtain

$$||c^k z - (c + (e-1))^k z|| \leq \sum_{j=1}^{k} \binom{k}{j} |\mu|^{k-j} ||f^j z - (f + (e-1))^j z||$$

for $k = 1,\ldots,n$. [Note that we could have used the binomial theorem to obtain this estimate because μ is in the center of A_1, but that we could not apply the binomial theorem in part (a) because e and f may not commute.]

Now there is a small positive ν such that $||f^j z - (f + (e-1))^j z|| < \nu$ for $j = 1,\ldots,n$ implies $||c^k z - (c + (e-1))^k z|| < \varepsilon$ for $k =$

$1,\ldots,n$ (for example, let $\nu = (1 + |\mu|)^{-n}\varepsilon$).

Using a computation similar to that in (a) of expanding $(f + (e - 1))^j$ and applying the norm inequalities we obtain that $||f^j z - (f + (e - 1))^j z||$ is less than or equal to a finite sum of numbers of the form

$$||f||^i ||e - 1||^{j-i-m-1} ||(e - 1)f^m z||$$

where $i, m = 0,1,\ldots,j-1$ with $i + m \leq j - 1$. The estimate $||e - 1|| \leq 1 + K$ enables us to choose $\eta > 0$ such that for $j = 1,\ldots,n$ we have $||f^j z - (f + (e - 1))^j z|| < \nu$. This completes the proof of the lemma. □

Remark. We could have used part (a) in the proof of (b) by replacing the condition $||(e - 1)z|| + ||(e - 1)f|| < \eta$ by the condition $||(e - 1)z|| + ||(e - 1)f|| + ||f(e - 1)|| < \eta$.

(19.3) Lemma. Let U be a bounded subset of the complex plane, and let $\varepsilon > 0$.

(a) If $c = f + \mu 1 \in A_1$ with $f \in A$, there is $\eta > 0$ such that if $e \in A$, $||e|| \leq K$, $||(e - 1)f|| + ||f(e - 1)|| < \eta$, and $t \in U$, then $||\exp(t(c + (e - 1))) - \exp(tc)|| \leq (\varepsilon + \exp(|t|(K + 1)) - 1)\exp(\mathrm{Re}\,(t\mu))$.

(b) If $c = f + \mu 1 \in A_1$ and $z \in X$, there is an $\eta > 0$ such that if $e \in A$ with $||e|| \leq K$, $||(e - 1)z|| + ||(e - 1)f|| < \eta$, and $t \in U$, then $||\exp(t(c + (e - 1)))z - \exp(tc)z|| < \varepsilon$.

Proof. (a) There exists an L such that $|t| \leq L$ for all $t \in U$. For each complex number t we have

$$||\exp(t(c + (e - 1))) - \exp(tc)|| = \exp \mathrm{Re}(t\mu)\,||\exp(t(f + (e - 1))) - \exp(tf)||.$$

Let us look at the factor $||\exp(t(f + (e - 1))) - \exp(tf)||$. If $t \in U$ and $e \in A$ with $||e|| \leq K$, then

$$||\exp(t(f + (e - 1))) - \exp(tf)|| \leq \sum_{k=1}^{\infty} \frac{|t|^k}{k!} ||(f + (e - 1))^k - f^k||$$

$$< \frac{\varepsilon}{3} + \sum_{k=1}^{n} \frac{|t|^k}{k!} ||(f + (e - 1))^k - f^k||$$

provided $\sum_{k=n+1}^{\infty} \frac{L^k}{k!} (||f|| + K + 1)^k < \frac{\varepsilon}{6}$. By Lemma (19.2),(a) we can choose $\eta > 0$ such that

$$\sum_{k=1}^{n} \frac{L^k}{k!} ||(f + (e - 1))^k - f^k - (e - 1)^k|| < \frac{\varepsilon}{3}$$

provided $||e|| \leq K$ and $||f(e - 1)|| + ||(e - 1)f|| < \eta$. If η and e are chosen like this, then

$$||\exp(t(f + (e - 1))) - \exp(tf)|| < \frac{2\varepsilon}{3} + \sum_{k=1}^{n} \frac{|t|^k}{k!} (K + 1)^k$$

$$< \varepsilon + \exp(|t|(K + 1)) - 1,$$

which proves (a).

(b) The proof of this is similar to part (a), and, in fact, is slightly simpler as the term $\exp |t|(K + 1)$ and factor $\exp \mathrm{Re}(\mu t)$ do not occur. It is a straightforward application of Lemma (19.2),(b), and is left to the reader. □

The exponential function in (19.3) may be replaced by any entire function provided that $\mu = 0$ in part (a).

§19. FACTORIZATION AND SEMIGROUPS

If $\beta > 0$, let

$$\Delta(\beta) = \{t \in C: |t| \leq \beta\}$$

and let

$$D(\beta) = \{t \in C: |t| \leq \beta \text{ and } \mathrm{Re}(t) \geq \beta^{-1}\}.$$

After an initial normalization and definition of a sequence required in the proof, the proof of Theorem (19.1) falls into two parts. We shall inductively choose a sequence $\{a_n\}$ in A_1 to satisfy certain conditions, and in the second part of the proof we use this sequence to define a^t, z_t, and y_t and check that the conclusions are satisfied.

Proof of Theorem (19.1):

We shall assume that $||z|| \leq 1$ and $||y|| \leq 1$, and that $\delta < 1$. We choose an increasing sequence of positive real numbers β_n that tend to infinity such that:

1°) $\Delta(\beta_1) \supset B$, $\beta_1 > 1$, and $1 + \exp m(K + 1) \leq \alpha_t$ for all $t \geq \beta_m$ and each positive integer m [the choice of the sequence $\{\beta_n\}$ with these properties is possible because $\alpha_t \to \infty$ as $t \to \infty$].

By induction we choose sequences $\{e_n\}_{n=1}^{\infty}$ and $\{a_n\}_{n=0}^{\infty}$ in A and A_1 respectively such that, for all positive integers n,

2°) $||e_n|| \leq K$;

3°) $a_o = 1$ and $a_n = \exp(\sum_{j=1}^{n} (e_j - 1))$;

4°) $||a_{n-1}^{-t}z - a_n^{-t}z|| \leq \delta\cdot 2^{-n}$ and $||ya_{n-1}^{-t} - ya_n^{-t}|| \leq \delta\cdot 2^{-n}$ for all $t \in \Delta(\beta_n)$; and

5°) $||a_{n-1}^{t} - a_n^{t}|| \leq \delta\cdot 2^{-n} + \{\exp |t|(K + 1) - 1\}\cdot\exp(-(n-1))\mathrm{Re}(t))$ for all $t \in \Delta(\beta_n)$.

We choose e_1 to satisfy the conclusions of Lemma (19.3) (a) and (b) simultaneously in the left and right module versions applied to the left and right Banach A-modules X and Y with $c = f = 0$, $\varepsilon = \delta/2$, $U = \Delta(\beta_1)$. This choice is possible using Lemma (18.2) because A has a bounded approximate identity bounded by K. Then e_1 and a_1 satisfy 2°) - 5°). Suppose $e_1,\ldots,e_n$ and $a_1,\ldots,a_n$ have been chosen to satisfy 2°) - 5°). We choose e_{n+1} to satisfy the conclusions of Lemma (19.3) (a) and (b) applied to the left and right Banach A-modules X and Y with $c = -n + f$, $f = \sum_{j=1}^{n} e_j$, $\varepsilon = \delta\cdot 2^{-n-1}\exp(-n\beta_{n+1})$, and $U = \Delta(\beta_{n+1})$. Because A has a bounded approximate identity bounded by K, we may choose e_{n+1} in A such that $||e_{n+1}|| \leq K$ and

$$||(e_{n+1} - 1)z|| + ||y(e_{n+1} - 1)|| + ||(e_{n+1} - 1)f|| + ||f(e_{n+1} - 1)|| < \eta,$$

where η is the smallest of the four values given by Lemma (19.3) with our initial conditions.

We now use conditions 2°) - 5°) to construct a^t, z_t, and y_t, and check that these elements satisfy the conclusions. By 4°) the sequence $\{a_n^{-t}z\}$ is Cauchy in X uniformly in t in $\Delta(\beta_m)$ for each positive integer m. Thus $\lim_{n\to\infty} a_n^{-t}z$ exists for all $t \in H$, and we denote this limit by z_t. Now the sequence $\{a_n^{-t}z\}$ converges to z_t uniformly in $t \in \Delta(\beta_m)$ for each positive integer m. Because $t \to a_n^{-t}z$ is analytic in H, $t \to z_t$ is analytic on Int $D(\beta_m)$ for each m. Hence $t \to z_t$ is analytic on H. The function $t \to y_t$ is obtained in a similar way. From 5°) we obtain $||a_{n-1}^t - a_n^t|| \leq \exp\{\beta_m(K + 1) - (n - 1)\beta_m^{-1}\} + \delta\cdot 2^{-n}$ for all $t \in D(\beta_m)$ and all $n \geq m$. Using this inequality as we used inequality 4°) above, we define $a^t = \lim_{n\to\infty} a_n^t$ for all $t \in H$, and obtain $t \to a^t$ is analytic.

We now check the conclusions of (19.1). The property $z = a^t z_t$ follows from the equation $z = a_n^t(a_n^{-t}z)$ and the definitions of a^t and z_t. The equalities $a^{t+s} = a^t a^s$ and $a^t z_{s+t} = z_s$ both follow from the corresponding results with a_n in place of a and $a_n^{-t}z$ in place of z_t. If $K = 1$ and $t \in R^+$, then

$$||a_n^t|| = ||\exp(t \sum_{j=1}^{n} (e_j - 1))|| \leq \exp(-nt)\cdot\exp(nt) = 1$$

for all positive integers n. In this case $||a^t|| \leq 1$ for all $t > 0$ proving (iv).

We now prove (v). Let $\varepsilon > 0$. There is a positive integer m such that $2^{-m+2}\cdot\delta < \varepsilon$. By 4°) we have $||a_{n-1}^t z - a_n^t z|| \leq \delta\cdot 2^{-n}$ for all $n \geq m$ and all $t \in \Delta(\beta_m)$ so that

$$||a^t z - a_m^t z|| \leq 2^{-m+1}\cdot\delta < \frac{\varepsilon}{2}$$

for all $t \in \Delta(1) \cap H \subset \Delta(\beta_m)$. Thus

$$\begin{aligned} ||a^t z - z|| &\leq \frac{\varepsilon}{2} + ||a_m^t z - z|| \\ &\leq \frac{\varepsilon}{2} + \sum_{j=1}^{\infty} \frac{|t|^j}{j!} m^j (K+1)^j \end{aligned}$$

for all $t \in \Delta(1) \cap H$. Hence there is a $\nu > 0$ such that $||a^t z - z|| < \varepsilon$ for all $t \in H$ with $|t| < \nu$. This completes the proof of (v).

By 5°) and the sum of a convergent geometric series, we have

$$||a^t|| \leq 1 + \sum_{n=1}^{\infty} ||a_{n-1}^t - a_n^t||$$

$$\leq 1 + \delta + \{\exp |t|(K + 1) - 1\}\cdot\{1 - \exp(-\mathrm{Re}(t))\}^{-1}$$

for $|t| \leq \beta_1$. The function of t on the right of this inequality has maximum

$$1 + \delta + (e^{K+1} - 1)(1 - e^{-1})^{-1}$$

on the interval $(0,1]$ because its derivative is positive. This proves (vi).

We choose β_1 so that $B \subset \Delta(\beta_1)$ (see p. 135). Thus

$$||z - z_t|| \leq \sum_{n=1}^{\infty} ||a_{n-1}^{-t}z - a_n^{-t}z|| \leq \delta$$

for $t \in B$ by 4°).

If $t \in H \cap \{\Delta(\beta_{m+1}) \setminus \Delta(\beta_m)\}$, then

$$||z_t|| \leq ||a_m^{-t}z|| + \sum_{j=m+1}^{\infty} ||a_{j-1}^{-t}z - a_j^{-t}z||$$

$$\leq ||a_m^{-t}z|| + 2^{-m}\cdot\delta$$

$$\leq \exp |t|(K + 1)m + 2^{-m}\cdot\delta$$

by 4°). Since $|t| \geq \beta_m$, by 1°) we have $1 + \exp m(K + 1) \leq \alpha_{|t|}$ so that $||z_t|| \leq \alpha_{|t|}^{|t|}$. This proves inequality (viii).

For each n and t, $a_n^{-t}z \in (Az)^-$ because $z \in (Az)^-$. Hence $z_t \in (Az)^-$ for all $t \in H$. This completes the proof of Theorem (19.1). □

§19. FACTORIZATION AND SEMIGROUPS

As an application of the preceding theorem we prove:

(19.4) Theorem. Let A be a separable Banach algebra with bounded two-sided approximate identity. Then there is an equivalent Banach algebra norm in which A has an abelian sequential bounded two-sided approximate identity of bound 1.

Proof. Let $\{z_n\}$ be a countable dense subset in A. Then there are elements z and y in A such that $z_n \in zA \cap Ay$ for all n (see Corollary (17.6)). [The elements z and y may also be obtained by applying (19.1) to the Banach A-bimodule $X = \{(a_n): a_n \in A, a_n \to 0$ as $n \to \infty\}$ in the usual way.] By Theorem (19.1), we obtain a function $t \to a^t$ for z and y in the Banach A-bimodule A. By (19.1) (v) and (vi), $a^t z \to z$ and $ya^t \to y$ as $t \to 0$, and $\{||a^t||: 0 < t \leq 1\}$ is bounded. This and $(zA)^- = A = (Ay)^-$ implies that $\{a^t: 0 < t \leq 1\}$ is an abelian bounded approximate identity for A. Let $m = \log ||a^1||$. Then $\{a^t e^{-mt}: t \in R^+\}$ is a bounded multiplicative semigroup in A, and as $t \to 0$ it is a bounded approximate identity for A since $e^{-mt} \to 1$ as $t \to 0$. A standard renorming of A gives an equivalent Banach algebra norm $||.||'$ on A with $||a^t||' \leq 1$ for all $t \in R^+$ [29, p. 18]. The fact that the approximate identity can be taken sequential follows from the proof of (1.5). □

The following strengthening of (19.4) was stated earlier as Theorem (3.5).

(19.5) Theorem. If A is a normed algebra with sequential bounded two-sided approximate identity $\{e_n\}$, then there is an equivalent norm in which A has a sequential bounded two-sided approximate identity $\{f_n\}$ of bound 1. Further, if A is complete, then $\{f_n\}$ may be chosen abelian.

Proof. In (19.4) the result was proved under the hypothesis that A is separable and complete. To obtain (19.5), let $\tilde{A}$ be the completion of A and let B be the closed subalgebra of $\tilde{A}$ generated by the set $\{e_n: n = 1,2,3,\ldots\}$. Then B is separable and complete, so there is a sequential abelian bounded two-sided approximate identity $\{f_n\}$ in B. But then it follows easily that $\{f_n\}$ is a bounded two-sided approximate identity for the whole of $\tilde{A}$. If A is incomplete, we choose $g_n \in A$ with $||g_n - f_n|| < 1/n$ $(n = 1,2,3,\ldots)$, and then $\{(1 - n^{-1})g_n\}$ is a sequential bounded two-sided approximate identity of bound 1 (but it may not be abelian!). This completes the proof. □

20. Analytic factorization I.

In this section we turn our attention to factoring with analytic functions. We know that if A is a Banach algebra with bounded approximate identity, then each $z \in A$ can be written $z = ay$ with $a, y \in A$. The element a may be chosen bounded by a fixed constant and y may be chosen arbitrarily close to z. In this setting the theorem below asserts that if f is a holomorphic function defined on a sufficiently large disc about $\xi = 1$, and satisfying $f(1) = 1$, then each $z \in A$ can be written $z = f(a)y$, where $a, y \in A$, with a bounded by a constant and y chosen close to a.

(20.1) Theorem. Let A be a Banach algebra with a left approximate identity $\{e_\lambda\}$ bounded by $K \geq 1$ and let X be a left Banach A-module. Let f be a holomorphic complex-valued function with $f(1) = 1$, defined on a neighborhood of $\{\xi \in C: |\xi - 1| \leq K + 1\}$.

Then for every $z \in X_e$ and $\delta > 0$ there exist elements $a \in A$ and $y \in X$ such that:

i) $z = f(a)y$;

ii) $||a|| \leq K$;

iii) $y \in \overline{Az}$;

iv) $||y - z|| < \delta$.

Proof. It suffices to prove the theorem in the case where f has no zeros in $(0,1)$, since otherwise we simply use the function

$$g(\xi) = f(e^{i\theta}\xi)f(e^{i\theta})^{-1}$$

for θ sufficiently small, instead of f. Let $\{\alpha_1,\ldots,\alpha_m\}$ denote the zeros of f in the disc $\{\xi \in C: |\xi - 1| < K + 1\}$ and write

$$f(\xi) = h(\xi)\cdot \prod_{i=1}^{m} (\xi - \alpha_i)^{k_i},$$

where h does not vanish on $\{\xi \in C: |\xi - 1| < K + 1\}$.

We divide the proof into four steps:

Step 1. Let $0 < c < 1$, $e_k \in \{e_\lambda\}$ $(k = 1,2,\ldots,n)$, and set $a_o = 1$,

$$a_n = (1 - c)^n + \sum_{k=1}^{n} c(1 - c)^{k-1}e_k.$$

Assume that no α_i belongs to the spectrum $\sigma(a_n)$ of a_n. Then $f(a_n)$

and $f(a_n)^{-1}$ are well defined elements in A_1 with

$$f(a_n)^{-1} = h(a_n)^{-1} \cdot \prod_{i=1}^{m} R(a_n,\alpha_i)^{k_i},$$

where $R(a_n,\alpha_i) = (a_n - \alpha_i)^{-1}$.

Proof. We assert first that $\sigma(a_n) \subset \{\xi \in C: |\xi - 1| < K + 1\}$. Indeed,

$$a_n - 1 = (1 - c)^n - 1 + \sum_{k=1}^{n} c(1 - c)^{k-1} e_k = \sum_{k=1}^{n} c(1 - c)^{k-1}(e_k - 1),$$

so that

$$||a_n - 1|| \leq (K + 1) \sum_{k=1}^{n} c(1 - c)^{k-1} = (K + 1)(1 - (1 - c)^n) < K + 1.$$

Then $h(a_n)^{-1}$ is a well defined element in A_1 since h does not vanish on $\{\xi \in C: |\xi - 1| < K + 1\}$. Finally,

$$f(a_n) = h(a_n) \cdot \prod_{i=1}^{m} (a_n - \alpha_i)^{k_i}$$

implies

$$f(a_n)^{-1} = h(a_n)^{-1} \cdot \prod_{i=1}^{m} R(a_n,\alpha_i)^{k_i}.$$

Step 2. If in addition a_n may be chosen so that

$$||f(a_{n+1})^{-1}z - f(a_n)^{-1}z|| < \frac{\delta}{2^{n+1}} \quad \text{for} \quad n = 0,1,2,\ldots,$$

then the theorem follows.

Proof. Set

$$y_n = f(a_n)^{-1}z, \quad n = 0,1,2,\ldots .$$

Then $\{y_n\}$ is a Cauchy sequence. With $y = \lim y_n$ we have $y \in \overline{Az}$ (see p. 96) and $||y - z|| < \delta$. Further, if

$$a = \sum_{k=1}^{\infty} c(1 - c)^{k-1}e_k,$$

then $||a|| \leq K$, $a = \lim a_n$, and $z = \lim f(a_n)y_n = f(a)y$ with $a \in A$.

The following technical step will be needed in the final induction step.

Step 3. There exists $c_o \in (0,1)$ such that

$$|c(1 - \tau)^{-1}| < \frac{1}{2(K+1)} \quad \text{for all } c \in (0,c_o] \text{ and all } \tau \in T_c,$$

where

$$T_c = \{\mu(1 - c)^{-n}: \mu \in \{\alpha_1,\ldots,\alpha_m\} \cup \{\xi \in C: |\xi - 1| = K + 1\}, n = 0,1,2,\ldots\}.$$

Let $p_\lambda = 1 - c + ce_\lambda$ for some $c \in (0,c_o]$. Then for $\tau \in T_c$ we have:

$R(p_\lambda,\tau)$ exists in A_1;

$||R(p_\lambda,\tau)|| \leq C < \infty$, where C depends only on f;

$\lim_\lambda R(p_\lambda,\tau)e = (1 - \tau)^{-1}e$ uniformly for $\tau \in T_c$ for fixed $e \in A$;

and

$\lim_\lambda R(p_\lambda,\tau)x = (1 - \tau)^{-1}x$ uniformly for $\tau \in T_c$ for fixed $x \in X_e$.

Proof. The first assertion is an immediate consequence of the fact that f has no zeros in $(0,1]$, so that

$$|1-\tau| \geq \gamma \quad \text{for all} \quad \tau \in T_c \quad \text{and all} \quad c \in (0,1),$$

where γ is some constant > 0.

Since

$$p_\lambda - \tau = (1-\tau)(1 - \frac{c}{1-\tau}(1-e_\lambda)),$$

the inverse $R(p_\lambda,\tau)$ exists in A_1 by the choice of c. Furthermore,

$$||R(p_\lambda,\tau)|| \leq |1-\tau|^{-1}(1 - \frac{1}{2(K+1)}(K+1))^{-1} = 2|1-\tau|^{-1}$$

$$\leq 2\max\{|1-\tau|^{-1}: \tau \in T_c\}.$$

Finally, the identity

$$R(p_\lambda,\tau) - (1-\tau)^{-1} = \frac{c}{1-\tau}R(p_\lambda,\tau)(1-e_\lambda)$$

shows that for fixed $e \in A$ and $x \in X_e$,

$$\lim_\lambda R(p_\lambda,\tau)e = (1-\tau)^{-1}e \quad \text{and} \quad \lim_\lambda R(p_\lambda,\tau)x = (1-\tau)^{-1}x$$

uniformly for $\tau \in T_c$.

Step 4. We will now by induction find a sequence $\{e_k\}$ with $e_k \in \{e_\lambda\}$ such that for any fixed $c \in (0,c_0]$, $R(a_n,\alpha_i)$ exists in A_1 for all $i = 1,\ldots,m$ and $n = 0,1,2,\ldots$, and

$$||f(a_{n+1})^{-1}z - f(a_n)^{-1}z|| < \frac{\delta}{2^{n+1}} \quad \text{for } n = 0,1,2,\ldots .$$

The theorem then follows from Step 2.

Proof. Suppose we have chosen $e_k \in \{e_\lambda\}$ $(k = 1,\dots,n)$ such that the $a_o, a_1, \dots, a_n$ have the above properties. Let $p_{n+1} = 1 - c + ce_{n+1}$, where $e_{n+1} \in \{e_\lambda\}$ is to be chosen.

Since $R(p_{n+1}, \tau)$ exists for $\tau \in T_c$, we may write

$$a_{n+1} - \alpha = (p_{n+1} - \frac{\alpha}{(1-c)^n})((1 - c)^n + \sum_{k=1}^{n} c(1 - c)^{k-1} R(p_{n+1}, \frac{\alpha}{(1-c)^n})e_k)$$

for $\alpha \in T_c$. We note that the second factor

$$a_n'(\alpha) = (1 - c)^n + \sum_{k=1}^{n} c(1 - c)^{k-1} R(p_{n+1}, \frac{\alpha}{(1-c)^n})e_k$$

may be chosen arbitrarily close to

$$(1 - c)^n + \sum_{k=1}^{n} c(1 - c)^{k-1}(1 - \frac{\alpha}{(1-c)^n})^{-1} e_k = (1 - \frac{\alpha}{(1-c)^n})^{-1}(a_n - \alpha)$$

uniformly for $\alpha \in T_c$ if we just take e_{n+1} large. Therefore, the inverse $a_n'(\alpha)^{-1}$ exists in A_1 and is uniformly bounded for $\alpha \in T_c$ and $e_{n+1} \in \{e_\lambda\}$ large. Hence also $R(a_{n+1}, \alpha) = (a_{n+1} - \alpha)^{-1}$ exists in A_1 and is uniformly bounded for $\alpha \in T_c$ and $e_{n+1} \in \{e_\lambda\}$ large. Since

$$f(a_{n+1})^{-1} = h(a_{n+1})^{-1} \cdot \prod_{i=1}^{m} R(a_{n+1}, \alpha_i)^{k_i}$$

it is only left to show that for each $x \in X_e$,

$$R(a_{n+1}, \alpha)x \to R(a_n, \alpha)x$$

uniformly for $\alpha \in T_c$ as e_{n+1} increases in $\{e_\lambda\}$. Now

$$R(a_{n+1},\alpha)x - R(a_n,\alpha)x = a_n'(\alpha)^{-1}R(p_{n+1},\frac{\alpha}{(1-c)^n})x - R(a_n,\alpha)x$$

$$= a_n'(\alpha)^{-1}[R(p_{n+1},\frac{\alpha}{(1-c)^n})x - (1-\frac{\alpha}{(1-c)^n})^{-1}x]$$

$$+ [(1-\frac{\alpha}{(1-c)^n})^{-1}a_n'(\alpha)^{-1} - R(a_n,\alpha)]x.$$

The first term can be made arbitrarily small by Step 3. The second term can be made arbitrarily small too, for we have already observed that

$$a_n'(\alpha) \to (1-\frac{\alpha}{(1-c)^n})^{-1}(a_n - \alpha) \quad \text{uniformly for } \alpha \in T_c,$$

so

$$a_n'(\alpha)^{-1} \to (1-\frac{\alpha}{(1-c)^n})R(a_n,\alpha) \quad \text{uniformly for } \alpha \in T_c.$$

The proof is now complete. □

21. Analytic factorization II.

We continue our study of factoring with analytic functions in this section. The motivating example for our main result is the following elementary factorization in the Banach algebra c_o of complex sequences converging to zero. If x is in c_o, then there is an $a \geq 0$ in c_o such that x is in $f(a)\cdot c_o$ for each continuous function f on $[0,\infty)$ with $f^{-1}(\{0\}) = \{0\}$ and f not tending to zero "too quickly"

§21. ANALYTIC FACTORIZATION II

as $t \to 0$. This observation is essentially a restatement of the obvious fact that for each sequence tending to zero there is a sequence tending to zero more slowly. We also observe that the element a has a functional calculus stronger than the analytic functional calculus. In this section a generalization of this result is proved. The algebra c_o is replaced by a Banach algebra with bounded approximate identity bounded by 1, and the functions that operate on a are taken to be analytic and bounded in the open disc $\Delta = \{z \in C: |z - 1/2| < 1/2\}$ and to converge to zero as $|z|$ tends to zero faster than some $|z|^{\alpha}$ $(\alpha > 0)$.

Preliminaries to the statement of Theorem (21.1).

If U is an open subset of the complex plane, then $\mathrm{Hol}(U)$ will denote the algebra of functions that are analytic in U. The crucial estimates that are needed in the proof of Theorem (21.1) are calculated in Lemma (21.2). We use the analytic functional calculus integrating around circles that contain the spectra of all the elements we are interested in. Throughout this section we write $f(a)$ for the element given by the operation on a in a Banach algebra by a function f analytic in a neighborhood of the spectrum of a. Use will be made of the following elementary estimate. If b is in a unital Banach algebra B, if Γ is a circle with center μ and radius r containing the spectrum of b, and if f is a function analytic in a neighborhood of Γ and its interior, then

$$||f(b)|| = ||\frac{1}{2\pi i}\int_{\Gamma} f(z)(z - b)^{-1}\,dz||$$

$$\leq r\cdot\sup\{|f(z)|: z \in \Gamma\}\cdot\sup\{||(z - b)^{-1}||: z \in \Gamma\}.$$

II. FACTORIZATION IN BANACH MODULES

In the proof of (21.1) we will take a function analytic in the open right half plane, and as the sequence $\{b_n\}$ is inductively chosen, a sequence of circles $\Gamma(n)$ is chosen so that the union of the interiors of $\Gamma(n)$ is the set $\{z \in C\colon \operatorname{Re} z > 1/2\}$. The $1/2$ can be replaced by $1 - \varepsilon$ by replacing $\mu_n - f_n$ as used in the proof of (21.1) by $\delta(\mu_n - f_n)$ for a sufficiently small positive δ.

We will assume that the Banach algebra A has a bounded two-sided approximate identity bounded by 1, i.e., for each finite subset F of A and each $\varepsilon > 0$ there is an $e \in A$ such that $||e|| \leq 1$ and $||ex - x|| + ||xe - x|| < \varepsilon$ for all $x \in F$. We restrict our attention to approximate identities bounded by 1 because of Theorem (3.5). Also, our proof will be given for left Banach modules only; minor modifications give the results for right Banach modules (all Banach modules are assumed to be isometric).

Let F be the algebra of functions in $\mathrm{Hol}(\Delta)$ such that for each f in F there is an $\alpha > 0$ with $|f(z)|\cdot|z|^{-\alpha}$ bounded on Δ. For each $\alpha > 0$, let

$$F_\alpha = \{f \in F\colon |f(z)|\cdot|z|^{-\alpha} \text{ is bounded on } \Delta\},$$

and, for each $f \in F_\alpha$, let

$$||f||_\alpha = \sup\{|f(z)|\cdot|z|^{-\alpha}\colon z \in \Delta\}.$$

The algebra F will operate on the element a that we shall construct, and the norm $||\cdot||_\alpha$ will occur in our inequalities.

Let β be a strictly increasing continuous real-valued function on

[0,1] with $\beta(0) = 0$. The only functions β that are of any interest to us are those which converge to zero faster than any power t^α as t tends to zero. Functions in F that converge to zero at zero more slowly than β will be those that give rise to factorization. Let

$$G(\beta) = \{g \in F\colon \text{there is a } K > 0 \text{ with } |g(z)| \geq K\beta(|z|) \text{ for all } z \in \Delta\}.$$

Even though $G(\beta)$ is not a multiplicative semigroup, there are multiplicative semigroups in $G(\beta)$ if β approaches zero very fast near zero. A useful example of such a semigroup is $\{w^\alpha : \alpha > 0\}$, where $w(z) = z$ for all $z \in \Delta$.

(21.1) <u>Theorem</u>. Let A be a Banach algebra with a bounded approximate identity Λ bounded by 1, let X be a left Banach A-module such that $AX = X$, and let Y be a right Banach A-module such that $YA = Y$. Let β be a continuous strictly increasing function from $[0,1]$ into $[0,1]$ such that $\beta(0) = 0$, and let $\varepsilon > 0$. If $x \in X$ and $y \in Y$, then (with the above notation) there are functions $\theta : F \to A$, $\chi : G(\beta) \to X$, and $\psi : G(\beta) \to Y$ such that:

i) θ is a homomorphism from F into A;

ii) if $g \in F$, then $\sigma(\theta(g)) \subset g(\Delta)^-$;

iii) if $w : \Delta \to \Delta\colon\ z \to z$, if $a = \theta(w)$, and if $g \in F$ has an analytic extension to a neighborhood of Δ^-, then $\theta(g) = g(a)$;

iv) for each $\alpha > 0$, the operator $\theta : (F_\alpha, ||\cdot||_\alpha) \to (A, ||.||)$ is continuous;

v) $x = \theta(g)\chi(g)$ and $y = \psi(g)\theta(g)$ for all $g \in G(\beta)$, and $\theta(g)\chi(gk) = \chi(k)$ for $g, k, gk \in G(\beta)$;

vi) $||\chi(g) - g(1/2)^{-1}\cdot x|| \le \varepsilon K(g)$ and $||\psi(g) - g(1/2)^{-1}\cdot y|| \le \varepsilon K(g)$ if $g \in G(\beta)$ and $K(g)\cdot|g(z)| \ge \beta(|z|)$ for all $z \in \Delta$;

vii) $||\theta(w^t)|| \le 1$ for all $t > 0$, and $||\theta(w^t)\cdot x - x|| + ||y\cdot\theta(w^t) - y|| \to 0$ as $t \to 0$ $(t > 0)$; and

viii) if Λ is closed and convex, and if $b \in \Lambda$ implies that $b^n \in \Lambda$ for $n = 1,2,\ldots$, then $\theta(w^t) \in \Lambda$ for all $t > 0$.

We shall need the following technical lemma to inductively choose the main sequence defined in the proof of Theorem (21.1). Throughout this section H will denote the open right half $\{z \in C\colon \operatorname{Re} z > 0\}$ of the complex plane C. If Γ is a subset of the complex plane and if G is a bounded function on Γ, then we let $||G||_\Gamma$ denote $\sup\{|G(z)|\colon z \in \Gamma\}$.

(21.2) Lemma. Let B be a unital Banach algebra, let X be a left Banach B-module, let $f \in B$, and let $\mu > ||f||$. If $\nu > 0$, then for each circle Γ with center $\mu + \nu$ and radius r, where $||f|| + \nu < r < \mu + \nu$, there are positive constants C_1, C_2, C_3, C_4 which depend on ν and r such that:

i) $||G(\mu-f+\nu(1-e))-G(\mu-f)-G(\mu+\nu(1-e))+G(\mu)|| \le \{C_1||(1-e)f|| + C_2||f(1-e)||\}\cdot||G||_\Gamma$ for all $G \in \operatorname{Hol}(H)$ and all $e \in B$ with $||e|| \le 1$;

ii) $||G(\mu-f+\nu(1-e))\cdot x-G(\mu-f)\cdot x|| \le \{C_1||(1-e)f||\cdot||x||+C_3||(1-e)x||\}||G||_\Gamma$ for all $G \in \operatorname{Hol}(H)$, all $x \in X$, and all $e \in B$ with $||e|| \le 1$; and

iii) if X is a Banach B-bimodule, then

$$||G(\mu-f+\nu(1-e))\cdot x-x\cdot G(\mu-f+\nu(1-e))+G(\mu-f)\cdot x-x\cdot G(\mu-f)|| \le$$
$$||G||_\Gamma\{2C_1||x||\cdot||(1-e)f||+2C_2||x||\cdot||f(1-e)||+C_4||ex - xe||\}$$

for all $G \in \operatorname{Hol}(H)$, all $x \in X$, and all $e \in B$ with $||e|| \le 1$.

§21. ANALYTIC FACTORIZATION II

Proof. Observe that Γ is contained in H, and that Γ surrounds the spectra of the elements $\mu - f + \nu(1 - e)$, $\mu - f$, and $\mu + \nu(1 - e)$ provided $||e|| \leq 1$. Throughout this proof we shall assume that $e \in B$ with $||e|| \leq 1$. The proof simply consists of estimating the required norms using the analytic functional calculus to evaluate $G(\mu-f+\nu(1-e))$ and the other terms. Our first estimate gives C_1 and is used in the three parts of the lemma. We shall regard Γ as a set or as a closed positively oriented path.

By the analytic functional calculus

$$G(\mu-f+\nu(1-e)) - G(\mu-f) = (2\pi i)^{-1} \int_\Gamma G(z)\{(z-\mu+f-\nu(1-e))^{-1}-(z-\mu+f)^{-1}\}dz$$

$$= (2\pi i)^{-1} \int_\Gamma G(z)(z-\mu+f-\nu(1-e))^{-1}\cdot\nu\cdot(1-e)\cdot(z-\mu+f)^{-1}dz.$$

In the last step we substitute

$$(z-\mu+f)^{-1} = (z-\mu)^{-1} - f(z-\mu)^{-1}(z-\mu+f)^{-1}$$

for all $z \in \Gamma$. We shall also use the inequalities

$$||(z-\mu+f-\nu(1-e))^{-1}|| \leq (|z-\mu-\nu|-||f+\nu e||)^{-1} \leq (r-||f||-\nu)^{-1}$$

and

$$||(z-\mu+f)^{-1}|| \leq (|z-\mu-\nu|-||f+\nu||)^{-1} \leq (r-||f||-\nu)^{-1}$$

for all $z \in \Gamma$.

Then (*)

$$||G(\mu-f+\nu(1-e))-G(\mu-f)-(2\pi i)^{-1} \int_\Gamma \nu G(z)(z-\mu+f-\nu(1-e))^{-1}(1-e)(z-\mu)^{-1}dz||$$

$$= \nu\cdot(2\pi)^{-1}|| \int_\Gamma G(z)(z-\mu+f-\nu(1-e))^{-1}(1-e)f(z-\mu)^{-1}(z-\mu+f)^{-1}dz||$$

$$\leq \nu\cdot(2\pi)^{-1}||G||_\Gamma(r-\nu-||f||)^{-2}||(1-e)f||(r-\nu)^{-1}2\pi r$$

$$= C_1||(1-e)f||\cdot||G||_\Gamma,$$

where $C_1 = \nu\cdot r\cdot(r-\nu)^{-1}(r-\nu-||f||)^{-2}$.

To prove i) we shall now compare the integral on the left hand side of (*) with $G(\mu+\nu(1-e))-G(\mu)$, and in doing this we require the following elementary rational equation. For $z \in \Gamma$,

$$(z-\mu)^{-1}(z-\mu+f-\nu(1-e))^{-1}\nu(1-e) - (z-\mu-\nu(1-e))^{-1} + (z-\mu)^{-1}$$

$$= (z-\mu)^{-1}(z-\mu+f-\nu(1-e))^{-1}\{(z-\mu-\nu(1-e))\nu(1-e)-(z-\mu)(z-\mu+f-\nu(1-e))$$
$$+ (z-\mu+f-\nu(1-e))\cdot(z-\mu-\nu(1-e))\}\cdot(z-\mu-\nu(1-e))^{-1}$$

$$= (z-\mu)^{-1}(z-\mu+f-\nu(1-e))^{-1}\cdot\nu\cdot f\cdot(e-1)(z-\mu-\nu(1-e))^{-1}.$$

This equation and the analytic functional calculus show that

$$||(2\pi i)^{-1}\int_\Gamma G(z)(z-\mu+f-\nu(1-e))^{-1}(1-e)(z-\mu)^{-1}dz-G(\mu+\nu(1-e))+G(\mu)||$$

$$= (2\pi)^{-1}||\int_\Gamma G(z)(z-\mu)^{-1}(z-\mu+f-\nu(1-e))^{-1}\cdot\nu\cdot f\cdot(e-1)(z-\mu-\nu(1-e))^{-1}dz||$$

$$\leq (2\pi)^{-1}||G||_\Gamma(r-\nu)^{-1}(r-\nu-||f||)^{-1}\cdot\nu\cdot||f(e-1)||(r-\nu)^{-1}\cdot 2\pi r$$

using estimates on the norms of the inverses similar to those used previously,

$$\leq C_2||G||_\Gamma||f(e-1)||,$$

where $C_2 = \nu(r-\nu)^{-2}(r-\nu-||f||)^{-1}r$. This shows that C_1 and C_2 exist for part i) for each circle with center $\mu + \nu$ and suitably restricted radius r.

We now show that C_3 exists. By (*), with all the elements operating on x, we have

$$||G(\mu-f-\nu(1-e))\cdot x-G(\mu-f)\cdot x||$$

$$\leq C_1||G||_\Gamma||(1-e)f||\cdot||x||+(2\pi)^{-1}||\int_\Gamma G(z)(z-\mu+f-\nu(1-e))^{-1}(z-\mu)^{-1}\nu(1-e)xdz||$$

$$\leq C_1||G||_\Gamma||(1-e)f||\cdot||x||+(2\pi)^{-1}||G||_\Gamma(r-||f||-\nu)^{-1}(r-\nu)^{-1}\nu||(1-e)x||2\pi r$$

$$= \{C_1||(1-e)f||\cdot||x||+C_3||(1-e)x||\}\cdot||G||_\Gamma$$

where $C_3 = \nu(r-||f||-\nu)^{-1}(r-\nu)^{-1}r$. We have used the bounds on the elements $(z-\mu+f-\nu(1-e))^{-1}$ and $(z-\mu)^{-1}$ obtained in proving part i) to complete the last inequality.

We now sketch the proof of part iii). By i) the left hand side L of iii) satisfies

$$L \leq \{2C_1||(1-e)f||+2C_2||f(1-e)||\}\cdot||x||\cdot||G||_\Gamma+||G(\mu+\nu(1-e))\cdot x-x\cdot G(\mu+\nu(1-e))||$$

because $G(\mu)\cdot x = x\cdot G(\mu)$.

Estimating the last term in this inequality we have

$$||G(\mu+\nu(1-e))\cdot x-x\cdot G(\mu+\nu(1-e))||$$

$$\leq (2\pi)^{-1}\cdot 2\pi r\cdot||G||_\Gamma\cdot\sup\{||(z-\mu-\nu(1-e))^{-1}\cdot x-x\cdot(z-\mu-\nu(1-e))^{-1}||: z \in \Gamma\}$$

$$\leq r\cdot||G||_\Gamma\cdot\nu\cdot||ex - xe||\cdot\sup\{||(z-\mu-\nu(1-e))^{-1}||^2: z \in \Gamma\}$$

$$\leq r\cdot\nu\cdot||G||_\Gamma\cdot(r-\nu)^{-2}||ex-xe||$$

which gives the required inequality if we choose $C_4 = r\cdot\nu\cdot(r-\nu)^{-2}$. This completes the proof of the lemma. □

Remark. To apply the above lemma to the open disc $\Delta = \{z \in C: |z - 1/2| < 1/2\}$ we map the disc conformally onto the open right half plane H and obtain most of the estimates using large circles like Γ contained in H.

Proof of Theorem (21.1):

Let $h(z) = (1+z)^{-1}$ for all $z \in H$. Then h is a conformal map from H onto the open disc Δ, and $h^{-1}(w) = w^{-1} - 1$ for all $w \in \Delta$. We shall use the map $g \to g\circ h : \mathrm{Hol}(\Delta) \to \mathrm{Hol}(H)$ to transform F to an algebra of functions in $\mathrm{Hol}(H)$ to which Lemma (21.2) applies.

The first step in the proof will be to make the inductive choice of the sequences used to define the functions θ and χ. Let $K > 1$ and let $\mu_n = K^n$ for $n = 0,1,2,\ldots$. Then

$$1 + (K-1)(1+K+\ldots+K^n) = \mu_{n+1} = \mu_n + (K-1)K^n \tag{1}$$

for $n = 0,1,2,\ldots$. We shall apply Lemma (21.2) to A_1, the Banach algebra obtained from A by adjoining an identity. By induction we choose sequences $\{e_n\}$ and $\{f_n\}$ in A such that

$$f_o = 0 \tag{2}$$

and

$$f_n = (K-1)(e_1+Ke_2+\ldots+K^{n-1}e_n);$$

$$||e_n|| \le 1, \tag{3}$$

and

$$||(1-e_{n+1})f_n||, \quad ||f_n(1-e_{n+1})||, \text{ and } ||(1-e_{n+1})x||$$

are so small that using the circle $\Gamma = \Gamma(n)$ with center $\mu_{n+1} = \mu_n + (K-1)K^n$ and radius $r = \mu_{n+1} - 1/2$ in Lemma (21.2) we obtain

$$C_1||(1-e_{n+1})f_n||+C_2||f_n(1-e_{n+1})|| \leq 2^{-n-2}\cdot\varepsilon \tag{4}$$

and

$$C_1||(1-e_{n+1})f_n||\cdot||x||+C_3||(1-e_{n+1})x|| \leq 2^{-n-1}\beta((2\mu_{n+1}+1)^{-1})\cdot\varepsilon \tag{5}$$

for $n = 0,1,2,\ldots$.

The inductive choice is possible because A has a bounded approximate identity bounded by 1. The definition of such an approximate identity ensures that e_{n+1} may be chosen with $||(1-e_{n+1})f_n||$ and $||f_n(1-e_{n+1})||$ very small. By Lemma (18.2) we may also choose e_{n+1} so that $||(1-e_{n+1})x||$ is very small since x is in $(AX)^-$.

In the second step of the proof we construct θ, by showing that $(g\circ h)(\mu_n-f_n)$ is a Cauchy sequence in A_1 for each $g \in F$, and we study some of the properties of θ. For convenience we shall write G for $g\circ h$ if $g \in F$. By Lemma (21.2)(i) with $f = f_n$, $\mu = \mu_n$, $\nu = (K-1)K^n$, $e = e_{n+1}$, $r = \mu + \nu - 1/2$ (note that $\mu_n \geq ||f_n|| + 1$), and by (4), we have: (6)

$$||G(\mu_{n+1}-f_{n+1})-G(\mu_n-f_n)|| \leq 2^{-n-2}\varepsilon||G||_{\Gamma(n)}+||G(\mu_{n+1}-(K-1)K^n e_{n+1})||+||G(\mu_n)||.$$

Now

$$||G||_{\Gamma(n)} = ||g||_{h^{-1}\circ\Gamma(n)} \leq ||g||_\Delta \leq \sup\{|g(z)|\cdot|z|^{-\alpha}: z \in \Delta\}$$

$$= ||g||_\alpha \text{ for all } g \in F_\alpha.$$

Estimating $|G(\mu_n)|$ we obtain

$$|G(\mu_n)| = |g((1+\mu_n)^{-1})| \leq ||g||_\alpha(1+\mu_n)^{-\alpha} = ||g||_\alpha(1+K^n)^{-\alpha}. \qquad (7)$$

To estimate the term $||G(\mu_{n+1}-(K-1)K^n e_{n+1})||$ we use the analytic functional calculus for the circle Γ with center $\mu_{n+1} = K^{n+1}$ and radius $R = (K-1)K^n + (1/2)\mu_n = \mu_{n+1} - (1/2)K^n$. Then

$$||G(\mu_{n+1}-(K-1)K^n e_{n+1})||$$

$$\leq (2\pi)^{-1}\cdot||G||_\Gamma\cdot\sup\{||(z-\mu_{n+1}+(K-1)K^n e_{n+1})^{-1}||: z \in \Gamma\}\cdot 2\pi R.$$

Further

$$\begin{aligned} ||G||_\Gamma &\leq \sup\{|z+1|^\alpha\cdot|g((1+z)^{-1})|\cdot|z+1|^{-\alpha}: z \in \Gamma\} \\ &\leq ||g||_\alpha\cdot\sup\{|z+1|^{-\alpha}: z \in \Gamma\} \\ &= ||g||_\alpha\cdot(1+\mu_{n+1}-R)^{-\alpha} \\ &= ||g||_\alpha\cdot(1+(1/2)K^n)^{-\alpha} \end{aligned}$$

where $\alpha > 0$ and $g \in F_\alpha$. Also

$$\begin{aligned} ||(z-\mu_{n+1}+(K-1)K^n e_{n+1})^{-1}|| &\leq (|z-\mu_{n+1}|-(K-1)K^n||e_{n+1}||)^{-1} \\ &\leq (R-(K-1)K^n)^{-1} = 2\cdot K^{-n} \quad \text{for all } z \in \Gamma. \end{aligned}$$

Using these two inequalities we have

$$\begin{aligned} ||G(\mu_{n+1}-(K-1)K^n e_{n+1})|| &\leq ||g||_\alpha(1+(1/2)K^n)^{-\alpha}\cdot R\cdot 2K^{-n} \\ &= ||g||_\alpha(1+(1/2)K^n)^{-\alpha}(2K-1) \end{aligned}$$

by definition of R. Substituting this estimate and (7) into (6) we have

$$||G(\mu_{n+1}-f_{n+1})-G(\mu_n-f_n)|| \le ||g||_\alpha\{2^{-n-2}\varepsilon+(2K-1)(1+\tfrac{1}{2}K^n)^{-\alpha}+(1+K^n)^{-\alpha}\} \quad (8)$$

$$\le ||g||_\alpha\{2^{-n-2}\cdot\varepsilon+(2K-1)2^\alpha K^{-n\alpha}+K^{-n\alpha}\}$$

for $n = 0,1,2,\ldots$, $\alpha > 0$, and $g \in F_\alpha$.

Since $K > 1$, inequality (8) shows that $\{G(\mu_n-f_n)\}$ is a Cauchy sequence in A_1 and thus converges to an element which we denote by $\theta(g)$. Since $g \to G(\mu_n-f_n)$ is a homomorphism from F into A_1 for each n, the mapping $\theta : F \to A_1$ is a homomorphism. If ϕ is the character on A_1 such that $\phi(a + \lambda 1) = \lambda$, then $\phi(G(\mu_n-f_n)) = G(\mu_n) \to 0$ as $n \to \infty$ for each $g \in F$ by (7). Hence θ is a homomorphism of F into A.

We now check properties ii) and iii). Assume $\alpha > 0$, $g \in F_\alpha$, and $\lambda \in C \setminus g(\Delta)^-$. Then $(\lambda 1-g)^{-1}$ is a bounded analytic function on Δ, and

$$|((\lambda 1-g)^{-1}-\lambda^{-1})(z)| = |\lambda|^{-1}\cdot|(\lambda 1-g)^{-1}(z)|\cdot|g(z)|$$

$$\le |\lambda|^{-1}\cdot||(\lambda 1-g)^{-1}||_\Delta\cdot||g||_\alpha\cdot|z|^\alpha$$

for all $z \in \Delta$ since $\lambda \neq 0$. Hence $(\lambda 1-g)^{-1} \in (F_\alpha)_1$ (the adjunction of an identity to F_α). The homomorphism $\theta : F_\alpha \to A$ defined above may be lifted to a homomorphism $\theta_1 : (F_\alpha)_1 \to A_1$ with $\theta_1|F_\alpha = \theta$ and $\theta_1(1) = 1$. Hence $(\lambda 1-\theta(g))^{-1} \in A_1$. This proves $\sigma(\theta(g)) \subset g(\Delta)^-$.

If $w : \Delta \to \Delta$: $t \to t$, we let $a = \theta(w)$. Then $\sigma(a) \subset \Delta^-$. Let $g \in F$ have an analytic extension g to a neighborhood of Δ^-. Now

$$\sigma(h(\mu_n-f_n)) = \sigma((1+\mu_n-f_n)^{-1}) \subset \Delta,$$

because $\sigma(\mu_n-f_n) = \mu_n-\sigma(f_n) \subset \mu_n+\{z \in C: |z| \le \mu_n-1\}$. By the continuity of the analytic functional calculus (see [29], for example) we have

$$g(a) = g(\lim(1+\mu_n-f_n)^{-1}) = \lim g((1+\mu_n-f_n)^{-1})$$

$$= \lim G(\mu_n-f_n) = \theta(g).$$

We now show that $\theta : (F_\alpha, ||.||_\alpha) \to (A, ||.||)$ is continuous. Let P denote the projection from A_1 onto A that annihilates the complex field C. Then

$$||\theta(g)|| = ||P\theta(g)|| \leq \sum_{n=0}^{\infty} ||PG(\mu_{n+1}-f_{n+1})-PG(\mu_n-f_n)||$$

$$\text{since } PG(\mu_0-f_0) = PG(\mu_0) = 0$$

$$\leq ||g||_\alpha \sum_{n=0}^{\infty} \{2^{-n-2}\varepsilon+(2K-1)\cdot 2^\alpha K^{-n\alpha}\}$$

from inequality (8) neglecting the final term $K^{-n\alpha}$ which does not arise because $G(\mu_n)$, from which it comes in (7), is annihilated by P. Thus

$$||\theta(g)|| \leq ||g||_\alpha\{\tfrac{\varepsilon}{2}+2^\alpha(2K-1)(1-K^{-\alpha})^{-1}\},$$

and θ is continuous on F_α.

In the final step of the proof we construct χ by showing that $\{G^{-1}(\mu_n-f_n)\cdot x\}$ is a Cauchy sequence in X for each $g \in G(\beta)$. By (5) and Lemma (21.2)(ii) we have

$$||G^{-1}(\mu_{n+1}-f_{n+1})\cdot x-G^{-1}(\mu_n-f_n)\cdot x|| \leq 2^{-n-1}\beta((2\mu_{n+1}+1)^{-1})\cdot||G^{-1}||_{\Gamma(n)}\cdot\varepsilon$$

where G^{-1} is the multiplicative inverse of G on H. Now

$$||G^{-1}||_{\Gamma(n)} \leq \sup\{|g((1+z)^{-1})|^{-1}: z \in \Gamma(n)\}$$

$$\leq K(g)\cdot\sup\{\beta(|(1+z)|^{-1})^{-1}: z \in \Gamma(n)\},$$

[where $K(g)$ is such that $K(g)\cdot|g(w)| \geq \beta(|w|)$ for all $w \in \Delta$ by definition of $G(\beta)$],

$$\leq K(g)\beta((1+\mu_{n+1}+\mu_{n+1}-(1/2))^{-1})^{-1} \leq K(g)\beta((1+2\mu_{n+1})^{-1})^{-1}$$

since β is an increasing function on $[0,1]$. Hence

$$||G^{-1}(\mu_{n+1}-f_{n+1})\cdot x - G^{-1}(\mu_n - f_n)\cdot x|| \leq 2^{-n-1}\cdot K(g)\cdot \varepsilon \tag{9}$$

for $n = 0,1,2,\ldots$ and $g \in G(\beta)$. Thus $\{G^{-1}(\mu_n - f_n)\cdot x\}$ is a Cauchy sequence in X and converges to an element which we denote by $\chi(g)$.

Since $x = G(\mu_n - f_n)\cdot G^{-1}(\mu_n - f_n)\cdot x$ for all n and $g \in G(\beta)$, we get $x = \theta(g)\chi(g)$ for all $g \in G(\beta)$. Similar reasoning shows that if $g \in F$ and k, $gk \in G(\beta)$, then $\theta(g)\chi(gk) = \chi(k)$. Also $G^{-1}(\mu_0 - f_0) = g((1+1)^{-1})^{-1}$ so that, by (9),

$$||\chi(g) - g(2^{-1})^{-1}x|| \leq \sum_{n=0}^{\infty} ||G^{-1}(\mu_{n+1}-f_{n+1})\cdot x - G^{-1}(\mu_n - f_n)\cdot x|| \leq \varepsilon\cdot K(g).$$

We now show that a bounded approximate identity for $(F, ||\cdot||_\Delta)$ acts as an approximate identity on x, and from this we obtain property vii). For each $g \in F$ and $n = 0,1,2,\ldots$

$$\begin{aligned} ||G(\mu_n - f_n)\cdot x - x|| &\leq ||x||\cdot||1 - G(\mu_n - f_n)|| \\ &\leq ||x||\cdot(2\pi)^{-1}\cdot 2\pi R\cdot||1-G||_\Gamma\cdot \sup\{||(z-\mu_n+f_n)^{-1}||: z \in \Gamma\}, \end{aligned} \tag{10}$$

[where Γ is the circle with center μ_n and radius $R = \mu_n - \frac{1}{2}$]

$$\leq ||x||\cdot R\cdot \sup\{|1-G(z)|: z \in \Gamma\}\cdot(R - ||f_n||)^{-1}$$

$$\leq ||x||\cdot K^n\cdot \sup\{|1-g(v)|: v \in \Delta,\ |v| \geq (2K^n+1)^{-1}\}\cdot 2$$

since $\{|1 - G(z)|: z \in \Gamma\} = \{|1 - g(v)|: v = 1/(1+z),\ z \in \Gamma\} \subset \{|1 - g(v)|: v \in \Delta$ and $|v| \geq (2K^n + 1)^{-1}\}$. By Lemma (21.2)(ii) and (5)

$$||G(\mu_{m+1}-f_{m+1})\cdot x-G(\mu_m-f_m)\cdot x|| \leq 2^{-m-1}\cdot\varepsilon\cdot\beta((2\mu_{m+1}+1)^{-1})\cdot||G||_{\Gamma(n)}$$

$$\leq 2^{-m-1}\cdot\varepsilon\cdot||g||_\Delta$$

since $\beta : [0,1] \to [0,1]$ and $||G||_{\Gamma(n)} \leq ||g||_\Delta$ (see the work after (6)). If $g \in F$, then

$$||\theta(g)\cdot x-x|| \leq ||G(\mu_n-f_n)\cdot x-x|| + \sum_{m=n}^{\infty} ||G(\mu_{m+1}-f_{m+1})\cdot x-G(\mu_m-f_m)\cdot x||$$

$$\leq 2||x||\cdot K^n\cdot\sup\{|1-g(v)|: v \in \Delta,\ |v| \geq (2K^n+1)^{-1}\} + 2^{-n}\cdot\varepsilon\cdot||g||_\Delta$$

by the previous two inequalities.

Choosing n very large so that the second term is small for $||g||_\Delta \leq 1$ (say) and then choosing $g \in F$ with $||g||_\Delta \leq 1$ and

$$\sup\{|1-g(v)|: v \in \Delta,\ |v| \geq (2K^n+1)^{-1}\}$$

very small, we see that a bounded approximate identity in the normed algebra $(F,||.||_\Delta)$ gives rise to an approximate identity for x in $\theta(F)$.

Clearly the functions $w^t : \Delta \to \Delta: z \to z^t$ for $t > 0$ are in F and $\{w^t: t > 0,\ t \to 0\}$ forms a bounded approximate identity in $(F,||.||_\Delta)$. Thus $||\theta(w^t)\cdot x-x|| \to 0$ as $t \to 0$. Finally, for $t > 0$ and $n = 1,2,...$ we have

$$||(1+\mu_n-f_n)^{-t}|| \leq (1+\mu_n)^{-t}\sum_{j=0}^{\infty}\left|\binom{-t}{j}\right|\cdot||f_n^j(1+\mu_n)^{-j}||$$

$$\leq (1+\mu_n)^{-t}\sum_{j=0}^{\infty}\binom{-t}{j}(-1)^j\left(\frac{\mu_n-1}{\mu_n+1}\right)^j = 2^{-t}$$

since $||f_n|| \leq \mu_n - 1$. Taking the limit as n tends to infinity we have $||\theta(w^t)|| \leq 2^{-t} \leq 1$ for all $t > 0$.

We now prove viii). For each positive integer n and each $t > 0$ we obtain

$$(1+\mu_n-f_n)^{-t} = (1+\mu_n)^{-t} \sum_{j=0}^{\infty} \binom{-t}{j} (-1)^j f_n^j (1+\mu_n)^{-j}$$

$$= (1+\mu_n)^{-t} + \sum_{j=1}^{\infty} \xi_{jn} (f_n(\mu_n-1)^{-1})^j$$

where $\xi_{jn} = \binom{-t}{j} (-1)^j \left(\frac{\mu_n-1}{\mu_n+1} \right)^j (1+\mu_n)^{-t}$. Now

$$f_n(\mu_n-1)^{-1} = \sum_{r=1}^{n} (K^n-1)^{-1}(K-1)K^{r-1} e_r \in \Lambda,$$

because $\Sigma_{r=1}^{n}(K^n-1)^{-1}(K-1)(K^{r-1}) = 1$, and hence $(f_n(\mu_n-1)^{-1})^j \in \Lambda$. Since $\Sigma_{j=1}^{\infty} \xi_{jn} = 2^{-t} - (1+\mu_n)^{-t}$, it follows that

$$(1+\mu_n-f_n)^{-t} \in (1+\mu_n)^{-t} + \{2^{-t}-(1+\mu_n)^{-t}\}\Lambda$$

for all $t > 0$. Letting n tend to infinity, we obtain $\theta(w^t) \in 2^{-t}\cdot\Lambda$ for all $t > 0$. The construction of ψ is similar to χ. □

22. Factorable Banach algebras without approximate units.

In this section we give examples of factorable Banach algebras which have neither left nor right approximate units.

Let B be any Banach algebra with identity, and suppose that $a \in B$, with $||a|| = 1$, is not a left zero divisor in B. Then $A = aB$ is an algebra and $||.||'$ defined by

$$||ab||' = ||b|| \quad \text{for} \quad b \in B$$

is a well defined Banach space norm on A. Since

$$||(ab_1)(ab_2)||' = ||b_1 ab_2|| \leq ||b_1|| \cdot ||b_2|| = ||ab_1||' ||ab_2||'$$

for $b_1, b_2 \in B$, this is an algebra norm as well.

(22.1) Proposition. Let B be a Banach algebra with identity, and suppose $a \in B$ with $||a|| = 1$ is a left topological zero divisor which is not a left zero divisor. Then the Banach algebra $A = aB$ with norm $||\cdot||'$ defined by $||ab||' = ||b||$ for $b \in B$ has neither left nor right approximate units.

Proof. Assume A has left approximate units. Then there exists an element u in B such that $||(au)a - a||' < 1$ and so $||ua - 1|| < 1$. Thus $(ua)^{-1}$ exists in B and $(ua)^{-1}ua = 1$. On the other hand, since a is a left topological zero divisor in B, a has no left inverse. Contradiction!

Now assume A has right approximate units. Then there exists an element u in B such that $||a(au) - a||' < 1$ and so $||au - 1|| < 1$. Thus $(au)^{-1}$ exists in B and $au(au)^{-1} = 1$. Hence $aB = B$. The open mapping theorem guarantees now that the continuous surjection

$$B \to B$$

$$b \to ab$$

is open. On the other hand, a is a left topological divisor of zero. Contradiction! Thus A has neither left nor right approximate units. □

§22. FACTORABLE BANACH ALGEBRAS WITHOUT APPROXIMATE UNITS

(22.2) Proposition. Let B be a Banach algebra with identity, and suppose $a \in B$ with $||a|| = 1$ is a left topological zero divisor which is not a left zero divisor. If there exist elements $b_1, b_2 \in B$ such that $b_1ab_2 = 1$, then $A = aB$ with norm $||\cdot||'$ defined by

$$||ab||' = ||b|| \quad \text{for } b \in B,$$

is a factorable Banach algebra which has neither left nor right approximate units.

Proof. Any element ab in A can be factored as

$$ab = (abb_1)(ab_2). \ \square$$

We give now an example of a situation in which the hypotheses of the preceding proposition are satisfied.

(22.3) Example. Let ℓ^1 be the Banach space of all absolutely summable sequences of complex numbers; normed in the usual way:

$$||x|| = \sum_{n=1}^{\infty} |x_n| < \infty \quad \text{for } x = (x_1, x_2, \ldots) \in \ell^1.$$

Let B be the Banach algebra of all bounded linear operators on ℓ^1 with operator-norm. Fix a sequence $\{\alpha_n\}$ of real numbers with $0 < \alpha_n \leq 1$ for $n = 1,2,\ldots$ and $\lim \alpha_n = 0$. We define operators $T, U, V \in B$ by

$$Tx = (x_1, \alpha_1 x_2, x_3, \alpha_2 x_4, x_5, \ldots);$$

$$Vx = (x_1, x_1, x_2, x_2, x_3, x_3, \ldots);$$

$$Ux = ((1+\alpha_1)^{-1}(x_1+x_2),(1+\alpha_2)^{-1}(x_3+x_4),\ldots).$$

We see immediately that $||T|| = 1$ and that T is one-to-one; hence T is not a left zero divisor in B. For $n = 1,2,\ldots$ let $E_n \in B$ be the orthogonal projection on the 2n-th coordinate. Then $||TE_n|| = \alpha_n$ and so $\lim ||TE_n|| = \lim \alpha_n = 0$. Since $||E_n|| = 1$, $n = 1,2,\ldots$, T is a left topological zero divisor in B. A direct computation shows that $UTV = I$. We now invoke the preceding proposition (with U, T, and V playing the role of b_1, a, and b_2 respectively) to see that TB, appropriately normed, is a Banach algebra all of whose elements can be factored, but which has neither left nor right approximate units.

(22.4) <u>Proposition</u>. Let B be a Banach *-algebra with identity, and suppose $a \in B$ with $||a|| = 1$ and $a^* = a$ is a left topological zero divisor which is not a left zero divisor. If there exists an element $b_o \in B$ such that $b_o^* a b_o = 1$, then $A = aB$ with involution $\#$ and norm $||\cdot||'$ defined by

$$(ab)^{\#} = a(b^*) \quad \text{and} \quad ||ab||' = ||b|| \quad \text{for } b \in B,$$

is a factorable Banach *-algebra which has neither left nor right approximate units.

<u>Proof</u>. Clearly, $\#$ is a well defined map from A into A. It is routine to check that $\#$ is an involution. Observe that $\#$ is isometric when * is. Indeed,

$$||(ab)^{\#}||' = ||a(b^*)||' = ||b^*|| = ||b|| = ||ab||' \quad \text{for } b \in B.$$

§22. FACTORABLE BANACH ALGEBRAS WITHOUT APPROXIMATE UNITS

Furthermore, it is interesting to notice that every positive functional on A is continuous as the following argument shows:

Let g be a positive functional on A. Define f on B by $f(b) = g(ab)$. Then f is a linear functional on B with

$$f(b^*b) = g(ab^*b) = g(ab^*b_o^*ab_ob)$$

$$= g(a(b_ob)^*ab_ob) = g((ab_ob)^{\#}(ab_ob)) \geq 0 \quad \text{for } b \in B;$$

i.e., f is a positive functional on B. Since B has an identity, f is continuous. Hence

$$|g(ab)| = |f(b)| \leq ||f||\cdot||b|| = ||f||\cdot||ab||' \quad \text{for } b \in B,$$

and so also g is continuous. □

We give now an example of a situation in which the hypotheses of the preceding proposition are satisfied.

(22.5) <u>Example</u>. Let ℓ^2 be the Hilbert space of all absolutely square-summable sequences of complex numbers with the usual inner product:

$$(x|y) = \sum_{n=1}^{\infty} x_n\overline{y}_n \quad \text{for } x = (x_1,x_2,\ldots) \text{ and } y = (y_1,y_2,\ldots) \in \ell^2.$$

Let B be the Banach *-algebra of all bounded linear operators on ℓ^2 with operator norm. Fix a sequence $\{\alpha_n\}$ of real numbers with $0 < \alpha_n \leq 1$ for $n = 1,2,\ldots$, and $\lim \alpha_n = 0$. Define $T \in B$ by

$$Tx = (x_1,\alpha_1x_2,x_3,\alpha_2x_4,x_5,\ldots).$$

Clearly, $||T|| = 1$, $T^* = T$, and T is one-to-one; hence T is not a left zero divisor. For $n = 1,2,\ldots$, let E_n be the orthogonal projection on the 2n-th coordinate. Then $||TE_n|| = \alpha_n$ and so $\lim ||TE_n|| = \lim \alpha_n = 0$. Since $||E_n|| = 1$, $n = 1,2,\ldots$, T is a left topological zero divisor in B. Now define $V \varepsilon B$ by

$$Vx = (x_1,0,x_2,0,x_3,0,\ldots).$$

Then

$$V^*x = (x_1,x_3,x_5,\ldots),$$

and direct computation shows that $V^*TV = I$.

We now invoke the preceding proposition (with T and V playing the role of a and b_o, respectively) to see that TB, appropriately normed, is a Banach algebra with isometric involution $\#$ all of whose elements can be factored, but which has neither left nor right approximate units.

We next give an example of a semisimple <u>commutative</u> Banach algebra A which factors but has no approximate units, i.e., every $a \varepsilon A$ can be written $a = bc$ with $b, c \varepsilon A$, but given $a \varepsilon A$ and $\varepsilon > 0$ there is no element $u \varepsilon A$ such that $||a - ua|| < \varepsilon$.

(22.6) <u>Example</u>. Let S be the semigroup, with pointwise addition, of all real sequences $b = \{b_n\}$, $n = 1,2,\ldots$, with $b_n > 0$ for almost all n, and let $A = \ell^1(S)$ be the corresponding convolution semigroup algebra (see [130, p.263]). A typical element $a \varepsilon A$ can be written

$$a = \sum_{n=1}^{\infty} \lambda_n \varepsilon_{f_n},$$

where ε_{f_n} is the Dirac measure on S concentrated at $f_n = \{f_{nk}\}_{k=1}^{\infty}$, and the norm on A is given by $||a|| = \Sigma_{n=1}^{\infty}|\lambda_n| < \infty$. Under convolution product, A is a commutative semisimple Banach algebra. For each positive integer m, let $H_m = \{f_{nm} : 1 \leq n \leq m,\ f_{nm} > 0\}$ and define

$$h_m = \begin{cases} \frac{1}{2} \min H_m, & \text{if } H_m \neq \emptyset, \\ 1, & \text{if } H_m = \emptyset. \end{cases}$$

The sequence $h = \{h_m\}$ is in S, and by construction we have $g_n = f_n - h$ contained in S for all $n = 1,2,\ldots$. This implies that

$$a = \varepsilon_h \cdot \left(\sum_{n=1}^{\infty} \lambda_n \varepsilon_{g_n}\right),$$

so A factorizes. On the other hand, if $g \in S$ and $a \in A$ we have $||\varepsilon_g \cdot a - \varepsilon_g|| \geq 1$, since g is not contained in the support of $\varepsilon_g \cdot a$. Hence A has no approximate units.

We remark that by introducing the weight function $w(b) = \min\{n : b_m > 0 \text{ for all } m, n\}$, where $b = \{b_n\} \in S$, and defining the norm by $||\Sigma_1^{\infty} \lambda_n \varepsilon_{f_n}|| = \Sigma_1^{\infty} |\lambda_n| w(f_n)$, we obtain a semisimple commutative Banach algebra which factorizes but whose norm is not equivalent to the norm on A induced by the regular representation; in particular A again does not have approximate units.

23. Nonfactorization.

We now briefly study some examples of nonfactorable Banach algebras with (necessarily unbounded) approximate identities. In this section G will denote a locally compact abelian group. Let $A^p(G)$, $1 \leq p < \infty$, be the linear space $L^1(G) \cap L^p(G)$ normed by $||f|| = ||f||_1 + ||f||_p$, $f \in A^p(G)$. Clearly if $p = 1$ (and G is arbitrary) or G is discrete (and

$1 \leq p < \infty$), then $A^p(G)$ is precisely the group algebra $L^1(G)$ and $||\cdot||$ is equivalent to $||\cdot||_1$. If G is compact, then $A^p(G) = L^p(G)$ and $||\cdot||$ is equivalent to $||\cdot||_p$.

(23.1) Lemma. $A^p(G)$, $1 \leq p < \infty$, with multiplication defined as convolution is a commutative Banach algebra with an approximate identity.

Proof. First let us show that $A^p(G)$ is complete. Suppose $\{f_n\}$ is a Cauchy sequence in $A^p(G)$. Since $L^1(G)$ and $L^p(G)$ are complete, there exist $f \in L^1(G)$ and $g \in L^p(G)$ such that $||f_n - f||_1 \to 0$ and $||f_n - g||_p \to 0$. Then there is a subsequence $\{f_{n_k}\}$ of $\{f_n\}$ such that $f_{n_k} \to f$ a.e. and a subsequence $f_{n_{k_\ell}}$ of it such that $f_{n_{k_\ell}} \to g$ a.e. Thus $f_{n_{k_\ell}} \to f$ a.e., so that $f = g$ a.e. Hence $A^p(G)$ is complete.

Now recall that $||f*g||_p \leq ||f||_1||g||_p$ for all $f \in L^1(G)$ and $g \in L^p(G)$, $1 \leq p < \infty$. Using this fact we immediately have for all $f, g \in A^p(G)$, $f*g \in A^p(G)$,

$$||f*g|| = ||f*g||_1 + ||f*g||_p \leq ||f||_1||g||_1 + ||f||_1||g||_p = ||f||_1||g||,$$

$$||f*g|| = ||f*g||_1 + ||f*g||_p \leq ||f||_1||g||_1 + ||f||_p||g||_1 = ||f||\cdot||g||_1,$$

and in particular,

$$||f*g|| \leq ||f||\cdot||g||.$$

Since G is commutative, also $A^p(G)$ is commutative. Finally, by the proof of Theorem (13.4), $A^p(G)$ has an approximate identity. This completes the proof. □

If $p = 1$, then $A^p(G) = L^1(G)$ has a bounded approximate identity by Theorem (13.4). If G is discrete, then $A^p(G) = L^1(G)$ even has an identity by Theorem (13.1). Thus if $p = 1$ (and G is arbitrary) or G is discrete (and $1 \leq p < \infty$), then $A^p(G)$ can be factored.

If $1 < p < \infty$ and G is nondiscrete, then $A^p(G)$ is a proper dense ideal in $L^1(G)$. Hence for $1 < p < \infty$ and G nondiscrete, $A^p(G)$ has no <u>bounded</u> approximate identity. Indeed, suppose $\{e_\lambda\}$ is an approximate identity in $A^p(G)$ bounded by K. Then for all $f \in A^p(G)$,

$$||f||_1 \leq ||f||_1 + ||f||_p = ||f||$$

$$= \lim ||e_\lambda f|| \leq \lim\sup ||e_\lambda||\cdot||f||_1 \leq K||f||_1.$$

But this means that $||\cdot||_1$ and $||\cdot||$ are equivalent on $A^p(G)$ which is a contradiction since $A^p(G)$ is a proper dense subset of $L^1(G)$.

Now the fundamental factorization theorem implies that $L^1(G)*A^p(G) = A^p(G)$ for all G and $1 \leq p < \infty$. Since the set $C_{oo}(G)$ of all continuous functions defined on G with compact support is dense both in $L^1(G)$ and $L^p(G)$, $C_{oo}(G)$ is dense in $A^p(G)$. Thus

$$A^p(G)*A^p(G)$$

is a dense subset of $A^p(G)$. For $p \geq 2$ and G nondiscrete, the linear subspace spanned by $A^p(G)*A^p(G)$ is a <u>proper</u> dense subset of $A^p(G)$ since the convolution of two L^2-functions is a C_o-function.

(23.2) Theorem. If $1 < p < \infty$ and G is nondiscrete, then $A^p(G)$ is a nonfactorable Banach algebra with a (necessarily unbounded) approximate identity.

24. Fréchet modules.

Let A be a Fréchet multiplicatively convex topological algebra (in short: Fréchet algebra). Let $\{||\cdot||_i\}_{i=1}^{\infty}$ be an increasing sequence of submultiplicative seminorms on A which determine the topology, i.e., if

$$U_i = \{a \in A: ||a||_i < i^{-1}\},$$

then $\{U_i\}_{i=1}^{\infty}$ is a countable base of absolutely convex idempotent neighborhoods of zero in A. Let A_i be the normed algebra with completion $\tilde{A}_i$ constructed from $||\cdot||_i$. We denote the norm on $\tilde{A}_i$ again by $||\cdot||_i$. Let $\pi_i: A \to \tilde{A}_i$ be the natural projection. For $j \geq i$ define $\pi_{ij}: A_j \to A_i$ by $\pi_{ij}(\pi_j(a)) = \pi_i(a)$. Since π_{ij} is a continuous homomorphism onto A_i it can be extended to a map from $\tilde{A}_j$ into $\tilde{A}_i$; we use the symbol $\tilde{\pi}_{ij}$ to denote this map. Note that since $U_i \subset U_j$ when $i \geq j$, the π_{ij} are norm decreasing. Thus a Fréchet algebra can be regarded as the projective limit of a sequence of Banach algebras.

We will often make use of the following fundamental result: Suppose $\{a_i\}_{i=1}^{\infty}$ is a sequence of elements such that $a_i \in \tilde{A}_i$ and $\tilde{\pi}_{ij}(a_j) = a_i$ whenever $j \geq i$; then there exists an element $a \in A$ such that $a_i = \pi_i(a)$ for all $i = 1,2,\ldots$.

(24.1) Definition. Let A be a Fréchet algebra. A left Fréchet

§24. FRÉCHET MODULES

A-module is a Fréchet space X which is also a left A-module in the algebraic sense such that

iv') $||ax||_i \leq M_i||a||_i||x||_i$ $(i = 1,2,\ldots)$

for all $a \in A$, $x \in X$, where M_i are constants ≥ 1, and $\{||\cdot||_i\}_{i=1}^{\infty}$ is an increasing sequence of seminorms on A, respectively X, such that if

$$U_i = \{a \in A\colon ||a||_i < i^{-1}\} \quad \text{and} \quad V_i = \{x \in X\colon ||x||_i < i^{-1}\}$$

then $\{U_i\}_{i=1}^{\infty}$ is a base of idempotent neighborhoods of zero in A and $\{V_i\}_{i=1}^{\infty}$ is a base of neighborhoods of zero in X.

(24.2) Remark. Clearly, condition (24.1.iv') implies that the bilinear mapping

$$A \times X \to X$$

$$(a,x) \to ax$$

is continuous.

(24.3) Remark. Let X be a left Fréchet A-module. Let A_i be the normed algebra with completion $\tilde{A}_i$ constructed from $||\cdot||_i$ on A and let X_i be the normed space with completion $\tilde{X}_i$ constructed from $||\cdot||_i$ on X. Let $\pi_i\colon A \to \tilde{A}_i$ and $\pi_i\colon X \to \tilde{X}_i$ be the natural projections. Then $\tilde{X}_i$ is in the obvious way a left Banach $\tilde{A}_i$-module such that

$$\pi_i(ax) = \pi_i(a)\pi_i(x) \quad \text{for all} \quad a \in A,\ x \in X.$$

For $j \geq i$ define $\pi_{ij}\colon A_j \to A_i$ by $\pi_{ij}(\pi_j(a)) = \pi_i(a)$ and

$\pi_{ij}:X_j \to X_i$ by $\pi_{ij}(\pi_j(x)) = \pi_i(x)$. Let $\tilde{\pi}_{ij}:\tilde{A}_j \to \tilde{A}_i$ and $\tilde{\pi}_{ij}:\tilde{X}_j \to \tilde{X}_i$ be their continuous extensions. Then

$$\tilde{\pi}_{ij}(ax) = \tilde{\pi}_{ij}(a)\tilde{\pi}_{ij}(x) \quad \text{for all} \quad a \in \tilde{A}_i,\ x \in \tilde{X}_i.$$

(24.4) Remark. Let A_1 be the Fréchet algebra obtained from A by adjoining an identity 1 with the usual topology defined by the semi-norms

$$||a + \alpha 1||_i = ||a||_i + |\alpha|.$$

If X is a left Fréchet A-module, then X is also a left Fréchet A_1-module, where

$$(a + \alpha 1)x = ax + \alpha x \quad \text{for} \quad a + \alpha 1 \in A_1 \quad \text{and} \quad x \in X.$$

Property (24.1.iv') holds for the same constants $M_i (i = 1,2,\ldots)$ because $M_i \geq 1$.

The mappings $\pi_i:A \to \tilde{A}_i$ and $\tilde{\pi}_{ij}:\tilde{A}_j \to \tilde{A}_i$ extend in the obvious way to the algebras obtained by adjoining an identity.

25. Essential Fréchet modules.

(25.1) Definition. Let A be a Fréchet algebra and let X be a left Fréchet A-module. Then the closed linear subspace of X spanned by

$$AX = \{ax: a \in A,\ x \in X\}$$

is called the essential part of X and is denoted by X_e. If $X = X_e$ then X is said to be an essential left Fréchet A-module.

§25. ESSENTIAL FRÉCHET MODULES

(25.2) <u>Definition</u>. A net $\{e_\lambda\}_{\lambda\varepsilon\Lambda}$ of elements in a Fréchet algebra is called a <u>bounded left approximate identity</u> if $\lim e_\lambda a = a$ for all $a \varepsilon A$ and $||e_\lambda||_i \leq K_i$ for all $\lambda \varepsilon \Lambda$ and $i = 1,2,\ldots,$ where K_i are constants ≥ 1 (depending only on i).

Observe that $\{\pi_i(e_\lambda)\}_{\lambda\varepsilon\Lambda}$ is a bounded left approximate identity in the normed algebra A_i and hence in $\tilde{A}_i$.

(25.3) <u>Proposition</u>. Let A be a Fréchet algebra with a bounded l.a.i. $\{e_\lambda\}_{\lambda\varepsilon\Lambda}$ and let X be a left Fréchet A-module. Then

$$X_e = \overline{AX} = \{x \varepsilon X: \lim e_\lambda x = x\}.$$

<u>Proof</u>. Suppose $x \varepsilon X_e$. Fix $i = 1,2,\ldots$. Then for any $\varepsilon > 0$ we can find $a_1,\ldots,a_n \varepsilon A$, $x_1,\ldots,x_n \varepsilon X$, such that

$$||\pi_i(x) - \sum_{k=1}^{n} \pi_i(a_k)\pi_i(x_k)||_i < \varepsilon.$$

Since

$$\lim_\lambda \pi_i(e_\lambda)(\sum_{k=1}^{n} \pi_i(a_k)\pi_i(x_k)) = \sum_{k=1}^{n} \lim_\lambda(\pi_i(e_\lambda)\pi_i(a_k))\pi_i(x_k)$$

$$= \sum_{k=1}^{n} \pi_i(a_k)\pi_i(x_k)$$

and $\{\pi_i(e_\lambda)\}_{\lambda\varepsilon\Lambda}$ is bounded in A_i, it follows that $\lim_\lambda \pi_i(e_\lambda x) = \lim_\lambda \pi_i(e_\lambda)\pi_i(x) = \pi_i(x)$. Thus, since i was arbitrary, $\lim_\lambda e_\lambda x = x$. □

(25.4) <u>Corollary</u>. Let A be a Fréchet algebra with a bounded l.a.i. $\{e_\lambda\}_{\lambda\varepsilon\Lambda}$. Then a left Fréchet A-module X is essential if and only if $\lim e_\lambda x = x$ for all $x \varepsilon X$.

(25.5) Remark. Let A be a Frechet algebra with a bounded left approximate identity. Then viewed as a left module over itself, A is an essential left Fréchet A-module.

(25.6) Proposition. Let A be a Fréchet algebra with a bounded left approximate identity. If X is a left Fréchet A-module, then the essential part X_e of X is a submodule of X, X_e is essential, and X_e contains any submodule of X which is essential.

26. Factorization in Fréchet modules.

(26.1) Definition. A net $\{e_\lambda\}_{\lambda\varepsilon\Lambda}$ of elements in a Fréchet algebra A is called a uniformly bounded left approximate identity if $\lim e_\lambda a = a$ for all $a \varepsilon A$ and $\sup_i ||e_\lambda||_i \leq K$ for all $\lambda \varepsilon \Lambda$, where K is some constant ≥ 1.

(26.2) Theorem. Let A be a Fréchet algebra with a left approximate identity uniformly bounded by $K \geq 1$ and let X be a left Fréchet A-module. Then for every $z \varepsilon X_e$, $j_o = 1,2,\ldots$, and neighborhood V of zero in X there exist elements $a \varepsilon A$ and $y \varepsilon X$ such that:

i) $z = ay$;

ii) $\sup_{i\leq j_o} ||a||_i \leq K$, (if $K = 1$, then $\sup_i ||a||_i \leq 1$);

iii) $y \varepsilon \overline{Az}$;

iv) $y - z \in V$.

In particular, $X_e = AX$.

Proof. Fix a real number $c > 0$ such that

$$0 < K < \frac{1-c}{c};$$

for example take

$$c = \frac{1}{K+2} \quad \text{or} \quad c = \frac{1}{2K+1}.$$

For $e \in A$ with $\sup_i ||e||_i \leq K$ define

$$\phi(e) = (1 - c + ce)^{-1}$$

Then $\phi(e)$ is a well defined element in A_1 since

$$1 - c + c\pi_i(e) = (1 - c)(1 + \frac{c}{1-c}\, \pi_i(e))$$

and

$$||1 - (1 + \frac{c}{1-c}\, \pi_i(e))||_i = \frac{c}{1-c}\, ||\pi_i(e)||_i \leq \frac{c}{1-c} K < 1 \quad \text{for all} \quad i = 1,2,\ldots$$

implies that $1 - c + c\pi_i(e)$ is invertible in $(\tilde{A}_i)_1$ for each $i = 1,2,\ldots$ (use the paragraph preceding (24.1)). The inequality

$$(1 - c - cK)||\phi(e)||_i \leq ||(1 - c)\phi(e)||_i - ||ce\phi(e)||_i$$

$$\leq ||(1 - c + ce)\phi(e)||_i = 1$$

shows that

$$||\phi(e)||_i \leq (1 - c - cK)^{-1} \quad \text{for all} \quad i = 1,2,\ldots .$$

We shall choose inductively a sequence $\{e_n\}$ of elements in A with $\sup_i ||e_n||_i \leq K$, $n = 1,2,\ldots$. Set

$$a_0 = 1,\ a_{n+1} = (1 - c + ce_{n+1})a_n,\quad n = 0,1,2,\ldots .$$

Then

$$a_n = (1 - c)^n + a_n',\ \text{where}\ \ a_n' \in A,$$

and

$$a_{n+1} - a_n = c(1 - c)^n(e_{n+1} - 1) + c(e_{n+1}a_n' - a_n').$$

Thus the infinite series

$$\sum_n ||a_{n+1} - a_n||_i$$

converges if the series

$$\sum_n ||e_{n+1}a_n' - a_n'||_i$$

converges.

Now set

$$y_n = a_n^{-1}z,\quad n = 0,1,2,\ldots .$$

Then

$$y_{n+1} - y_n = ca_{n+1}^{-1}(z - e_{n+1}z).$$

Since

$$a_{n+1}^{-1} = \phi(e_1)\phi(e_2)\cdots\phi(e_{n+1}),$$

$$||a_{n+1}^{-1}||_i \leq (1 - c - cK)^{-(n+1)},$$

and so

$$\sum_{n=0}^{\infty} ||y_{n+1} - y_n||_i \leq \sum_{n=0}^{\infty} cM_i(1 - c - cK)^{-(n+1)}||e_{n+1}z - z||_i.$$

Now, since $V_i = \{x \in X: ||x||_i < i^{-1}\}$, $i = 1,2,\ldots$, is a decreasing fundamental sequence of neighborhoods of zero in X, there exists an $i_o \geq j_o$ such that $V_{i_o} \subset V$. Fix δ, $0 < \delta < i_o^{-1}$. We choose inductively e_{n+1} in A with $\sup_i ||e_{n+1}||_i \leq K$ such that

$$||e_{n+1}a_n' - a_n'||_{i_o+n} < \varepsilon_{n+1},$$

and

$$||e_{n+1}z - z||_{i_o+n} < \varepsilon_{n+1},$$

where

$$\varepsilon_{n+1} = \frac{1}{cM_{i_o+n}}(1 - c - cK)^{n+1}\frac{\delta}{2^{n+1}}.$$

Then

$$\sum_{\substack{n \geq i-i_o \\ n \geq 0}} ||y_{n+1} - y_n||_i \leq \sum_{\substack{n \geq i-i_o \\ n \geq 0}} ||y_{n+1} - y_n||_{i_o+n}$$

$$\leq \sum_{\substack{n \geq i-i_o \\ n \geq 0}} cM_{i_o+n}(1 - c - cK)^{-(n+1)}||e_{n+1}z - z||_{i_o+n}$$

$$< \sum_{\substack{n \geq i-i_o \\ n \geq 0}} \frac{\delta}{2^{n+1}} \leq \delta .$$

Also $\{\pi_i(a_n)\}_n$ is a Cauchy sequence in $(\tilde{A}_i)_1$ since

$$\sum_{\substack{n \geq i-i_o \\ n \geq 0}} ||e_{n+1}a'_n - a'_n||_i \leq \sum_{\substack{n \geq i-i_o \\ n \geq 0}} ||e_{n+1}a'_n - a'_n||_{i_o+n}$$

$$\leq \sum_{\substack{n \geq i-i_o \\ n \geq 0}} \frac{1}{cM_{i_o+n}}(1 - c - cK)^{n+1}\frac{\delta}{2^{n+1}} \leq \frac{\delta}{c}.$$

If $\tilde{a}_i = \lim_n \pi_i(a_n)$, then $\tilde{a}_i \in \tilde{A}_i$ since $\tilde{a}_i = \lim_n \pi_i(a'_n)$. Clearly $\tilde{\pi}_{ij}(\tilde{a}_j) = \tilde{a}_i$ whenever $j \geq i$; hence there exists an element $a \in A$ such that $\tilde{a}_i = \pi_i(a)$ for all $i = 1,2,\ldots$. Since $\pi_i(z) = \lim_n \pi_i(a_n)\pi_i(y_n) = \tilde{a}_i\tilde{y}_i = \pi_i(a)\pi_i(y) = \pi_i(ay)$ for all $i = 1,2,\ldots$, it follows that $z = ay$.

The inequality

$$||a_n||_i \leq (1 - c + cK)^n, \quad n = 1,2,\ldots$$

gives for $K = 1$ the precise estimate

$$||a||_i \leq K = 1 \quad \text{for all} \quad i = 1,2,\ldots .$$

For arbitrary K we observe that

$$a = a_o + \sum_{n=0}^{\infty} (a_{n+1} - a_n)$$

$$= \sum_{n=0}^{\infty} c(1 - c)^n e_{n+1} + c \sum_{n=0}^{\infty} (e_{n+1}a'_n - a'_n)$$

implies:

$$||a||_{i_o} \leq K + \delta.$$

Thus we have shown the existence of a factorization $z = ay$ with $||a||_{i_o} \leq K + \delta$ and $||y - z||_{i_o} \leq \delta$, where δ can be arbitrarily small. Let α be any number with $0 < \alpha < 1$ and consider the factorization $z = (\alpha a)(\alpha^{-1}y)$. If δ, $0 < \delta < i_o^{-1}$, is sufficiently small and α sufficiently near by 1, then $||\alpha a||_{i_o} \leq K$ and at the same time $||\alpha^{-1}y - z||_{i_o} \leq 2\delta$. □

Interesting examples of Fréchet algebras with uniformly bounded approximate identities may be constructed from the so-called Beurling algebras.

(26.3) <u>Example</u>. Let

$$\omega_o(x) = 1 + |x|, \quad \omega_n(x) = (1 + |x|)^{1-\frac{1}{n}}, \quad x \in R,$$

and consider the Banach algebras

$$A_n = \{f \in L^1(R): ||f||_n \equiv \int_{-\infty}^{+\infty} |f(x)|\omega_n(x)\, dx < \infty\}, \quad n = 0,1,2,\ldots$$

under convolution multiplication and norm $||\cdot||_n$. Then it is easy to see that

$$A = \bigcap_{n=1}^{\infty} A_n$$

topologized by the increasing sequence of norms $||\cdot||_n$, $n = 1,2,\ldots$, is a Fréchet algebra.

To construct an approximate identity, take any sequence $\{u_k\}_{k=1}^{\infty}$ of nonnegative functions in $L^1(R)$ with $\text{supp } u_k \subset [-\frac{1}{k},\frac{1}{k}]$ and

$$\int_{-\infty}^{+\infty} |u_k(x)| \, dx = 1.$$

Then $\{u_k\}_{k=1}^{\infty}$ is an approximate identity in $L^1(R)$. Since each function u_k has compact support, it follows that $\{u_k\}_{k=1}^{\infty}$ is in particular an approximate identity in A_n for each $n = 0,1,2,\ldots$. Thus $\{u_k\}_{k=1}^{\infty}$ is an approximate identity in A. Now

$$||u_k||_o = \int_{-\infty}^{+\infty} u_k(x)(1 + |x|) \, dx \leq 2 \int_{-1}^{+1} u_k(x) \, dx \leq 2,$$

and $||u_k||_n \leq ||u_k||_o$ for $n = 1,2,\ldots$. Hence the approximate identity $\{u_k\}_{k=1}^{\infty}$ in A is uniformly bounded by $K = 2$ and so the factorization theorem applies to this algebra.

It seems worthwhile to state the factorization theorem for the special case of a Fréchet A-module over a Banach algebra.

(26.4) Theorem. Let A be a Banach algebra with a left approximate identity bounded by $K \geq 1$ and let X be a left Fréchet A-module. Then for every $z \in X_e$ and neighborhood V of zero in X there exist elements $a \in A$ and $y \in X$ such that:

i) $z = ay$;

ii) $||a|| \leq K$;

iii) $y \in \overline{Az}$;

iv) $y - z \in V$.

In particular, $X_e = AX$.

It would be natural to ask if the version of Theorem (26.4) would hold when X is only assumed to be a complete locally convex space. As the next example shows this is not the case.

(26.5) Example. Let S be a locally compact Hausdorff space which is not compact and let $C_o(S)$ be the Banach algebra of all complex valued continuous functions on S which vanish at infinity with sup-norm $||\cdot||_\infty$. Consider the space $C(S)$ of all complex-valued bounded continuous functions on S endowed with the strict topology β. The strict topology β is the locally convex topology defined on $C(S)$ by the family of seminorms:

$$||f||_\phi = ||\phi f||_\infty, \quad \phi \in C_o(S).$$

One can show that $C(S)$ endowed with the strict topology is a non-metrizable complete locally convex space. Obviously, $C(S)$ is a $C_o(S)$-module with respect to pointwise multiplication and

$$||\psi f||_\phi \leq ||\psi||_\infty ||f||_\phi \quad \text{for all} \quad \phi,\psi \in C_o(S) \quad \text{and} \quad f \in C(S).$$

Furthermore, $C_o(S)$ has a bounded approximate identity $\{e_\lambda\}$, and $\beta\text{-}\lim_\lambda e_\lambda f = f$ for all $f \in C(S)$. It is clear however that the function $f(x) = 1$ $(x \in S)$ in $C(S)$ cannot be factored as fg, with $f \in C_o(S)$ and $g \in C(S)$.

CHAPTER III

MORE ABOUT APPROXIMATE IDENTITIES

27. Local versus global.

Let A be a *-algebra. An element $x \varepsilon A$ is said to be selfadjoint if $x^* = x$. If A has a left identity 1, then

$$1^* = 11^* = (11^*)^* = 1^{**} = 1;$$

thus 1 is a selfadjoint two-sided identity in A.

An element $x \varepsilon A$ is said to be normal if $x^*x = xx^*$. Any normal element $x \varepsilon A$ is, by Zorn's Lemma, contained in a maximal commutative *-subalgebra of A.

Obviously, if A has an identity (a global property) then every maximal commutative *-subalgebra of A has an identity (a local property). What about the converse?

(27.1) Proposition. Let A be a *-algebra which has no nonzero nilpotent selfadjoint elements. If every maximal commutative *-subalgebra has an identity, then A has an identity.

Proof. Let B and B' be maximal commutative *-subalgebras of A. Let e, e' be the identities of B, B' respectively. We want to show that e = e'. Observe that

$$e(e' - e'e - ee' + ee'e) = 0,$$

and

$$(e - ee' - e'e + e'ee')e' = 0.$$

Since there exists no nonzero selfadjoint element annihilating a maximal commutative *-subalgebra with identity we obtain

$$(1) \qquad e' = e'e + ee' - ee'e,$$

$$(2) \qquad e = ee' + e'e - e'ee'.$$

From (1) we have

$$(3) \qquad e' = e'e'e' = 2e'ee' - e'ee'ee'$$

and so

$$(4) \qquad ee'e = 2ee'ee'e - ee'ee'ee'e.$$

Set $u = ee'e$. Clearly u is selfadjoint and (4) implies $(u - u^2)^2 = 0$. Hence by assumption we have $u = u^2$. Similarly $e'ee'$ is an idempotent. Then from (3) $e' = e'ee'$, and so from (2) $e + e' - ee' - e'e = 0$; thus $(e - e')^2 = 0$, which implies by assumption that $e = e'$.

Hence all maximal commutative *-subalgebras of A have the same identity e. We show that e is an identity in A. Since every element $x \in A$ has a (unique) decomposition of the form $x = h + ik$ with h, k selfadjoint it is sufficient to show that $he = eh = h$ for every selfadjoint h in A. But this follows from the existence of a maximal commutative *-subalgebra containing h. Thus the proof is complete. □

(27.2) Corollary. A C^*-algebra A has an identity if and only if every maximal commutative *-subalgebra of A has an identity.

Proof. In a C^*-algebra we have $||h^{2^n}|| = ||h||^{2^n}$, $n = 1,2,\ldots$, for all selfadjoint elements h. Thus A has no nonzero nilpotent selfadjoint elements and the preceding proposition applies. □

28. Well-behaved approximate identities.

Let S be a locally compact Hausdorff space. We showed in (12.1) that S is compact if and only if the algebra $C_o(S)$ of continuous complex-valued functions which vanish at infinity has an identity. In this section we shall study a special type of approximate identity which, among other things, can be used to characterize paracompactness in an analogous fashion.

(28.1) Definition. An approximate identity $\{e_\lambda\}_{\lambda\in\Lambda}$ in a C^*-algebra A is said to be canonical if $e_\lambda \geq 0$ for all $\lambda \in \Lambda$ and $e_\mu e_\lambda = e_\mu$ for $\mu < \lambda$. A well-behaved approximate identity (w.a.i.) in A is a canonical approximate identity $\{e_\lambda\}_{\lambda\in\Lambda}$ such that if $\lambda \in \Lambda$ and $\{\lambda_n\}$ is a strictly increasing sequence in Λ, then there is a positive integer n_o such that $e_\lambda e_{\lambda_{n_1}} = e_\lambda e_{\lambda_{n_2}}$ for $n_1, n_2 \geq n_o$.

(28.2) Example. Let S be the real line R with the usual topology. We shall construct a well-behaved approximate identity for $C_o(S)$. Let Λ denote the set of pairs (i,j), where i is any positive integer and

$j = 0$ or $j = 1$. Order Λ as follows:

(1) $(i,j) = (i',j')$ if $i = i'$ and $j = j'$;

(2) $(j,0) > (i,1)$ for all integers i and j;

(3) $(i,0) > (j,0)$ if $i > j$.

If $\lambda = (i,0)$, choose f_λ in $C_o(S)$ such that $0 \leq f_\lambda \leq 1$ and $f_\lambda \equiv 1$ on $[-i,i]$ and $f_\lambda \equiv 0$ off $[-(i+1),(i+1)]$.

If $\lambda = (i,1)$, choose f_λ in $C_o(S)$ such that $0 \leq f_\lambda \leq 1$, $f_\lambda(x_i) = 1$, where $x_i = \frac{1}{2}(1/(i+1) + 1/i)$, and $f_\lambda \equiv 0$ off $[1/(i+1),1/i]$.

Then it is easily checked that the net $\{f_\lambda\}_{\lambda\varepsilon\Lambda}$ is a well-behaved approximate identity for $C_o(S)$.

(28.3) <u>Example</u>. Let A be the C^*-algebra of compact operators on a Hilbert space H, let $\{e_\gamma\}_{\gamma\varepsilon\Gamma}$ be an orthonormal basis for H, and let Λ be the set of all finite subsets of Γ ordered by inclusion. If $\lambda \varepsilon \Lambda$, let P_λ be the finite-dimensional projection on H defined by

$$P_\lambda \xi = \sum_{\gamma\varepsilon\lambda} (\xi|e_\gamma)e_\gamma$$

for $\xi \varepsilon H$, where $(.|.)$ is the inner product on H. Then $\{P_\lambda\}_{\lambda\varepsilon\Lambda}$ is a well-behaved approximate identity for A.

(28.4) <u>Proposition</u>. Let A be a C^*-algebra. If $\{e_n\}$ is a sequential canonical approximate identity in A, then $\{e_n\}$ is well-behaved.

<u>Proof</u>. Let m and $n_1 < n_2 < \ldots$ be positive integers. Choose a positive integer i_o so that $n_i > m$ for $i \geq i_o$. Then $e_m e_{n_{i_1}} = e_m = e_m e_{n_{i_2}}$ for $i_1, i_2 \geq i_o$ by the canonical property; so $\{e_n\}$ is well-behaved. □

(28.5) Proposition. Let $\{A_\gamma\}_{\gamma\in\Gamma}$ be a family of C^*-algebras. If each A_γ has a well-behaved approximate identity, then the subdirect sum $(\sum_{\gamma\in\Gamma} A_\gamma)_o$ has a well-behaved approximate identity.

Proof. Recall that the subdirect sum $(\sum_{\gamma\in\Gamma} A_\gamma)_o$ of the family $\{A_\gamma\}_{\gamma\in\Gamma}$ is the class of all functions $a = (a(\gamma))$ defined on Γ with $a(\gamma) \in A_\gamma$ for each $\gamma \in \Gamma$ such that $\{\gamma \in \Gamma: ||a(\gamma)|| \geq \varepsilon\}$ is finite for each $\varepsilon > 0$. The algebraic operations are pointwise and

$$||a|| = \sup\{||a(\gamma)||: \gamma \in \Gamma\}.$$

Now for each $\gamma \in \Gamma$ let $\{e_{\gamma,\lambda_\gamma}\}_{\lambda_\gamma\in\Lambda_\gamma}$ be a well-behaved approximate identity in A_γ. Let Λ be the class of all functions λ whose domain D_λ is a finite subset of Γ with $\lambda(\gamma) \in \Lambda_\gamma$ for each $\gamma \in D_\lambda$. Λ is a directed set with respect to the partial ordering $\leq$ defined by: $\lambda_1 \leq \lambda_2$ iff $D_{\lambda_1} \subset D_{\lambda_2}$ and $\lambda_1(\gamma) \leq \lambda_2(\gamma)$ for each $\gamma \in D_{\lambda_1}$. For each $\lambda \in \Lambda$ define e_λ in $(\sum_{\gamma\in\Gamma} A_\gamma)_o$ by $e_\lambda(\gamma) = e_{\gamma,\lambda(\gamma)}$ for each $\gamma \in D_\lambda$ and $e_\lambda(\gamma) = 0$ otherwise. It is straightforward to verify that $\{e_\lambda\}_{\lambda\in\Lambda}$ is a well-behaved approximate identity in $(\sum_{\gamma\in\Gamma} A_\gamma)_o$. □

(28.6) Example. If S is a locally compact paracompact Hausdorff space then the commutative C^*-algebra $C_o(S)$ has a well-behaved approximate identity bounded by one.

Proof. Since S is paracompact, S is the union of a collection $\{S_\gamma\}_{\gamma\in\Gamma}$ of nonempty pairwise disjoint clopen σ-compact subsets of S. Since each C^*-algebra $C_o(S_\gamma)$ has a sequential canonical approximate identity bounded by one (see the proof of Proposition (12.2)) and since

$C_o(S) \simeq (\sum_{\gamma\varepsilon\Gamma} C_o(S_\gamma))_o$, it follows by (28.5) that $C_o(S)$ has a w.a.i. bounded by one. □

(28.7) Proposition. If a normed algebra A has a countable a.i. $\{u_n\}$ and $\{e_\lambda\}_{\lambda\varepsilon\Lambda}$ is another a.i. for A, then there is a countable subset Λ_o of Λ so that $\{e_\lambda\}_{\lambda\varepsilon\Lambda_o}$ is an a.i. for A.

Proof. Choose a countable subset Λ_o of Λ such that $\lim_{\lambda\varepsilon\Lambda_o} e\, u_n = u_n = \lim_{\lambda\varepsilon\Lambda_o} u_n e_\lambda$ for each n. Then $\{e_\lambda\}_{\lambda\varepsilon\Lambda_o}$ is an a.i. for A. □

(28.8) Theorem. Let A be a C^*-algebra with a countable approximate identity. Then A has a well-behaved approximate identity.

Proof. Let $\{c_n\}$ be a countable a.i. for A. We may assume that $c_n \geq 0$ since $c_n^* c_n$ is also an approximate identity for A (to see this note that

$$\begin{aligned} ||c_n^* c_n a - a|| &\leq ||c_n^*(c_n a - a)|| + ||c_n^* a - a|| \\ &\leq ||R_{c_n}||\cdot||(c_n a - a)^*|| + ||(a^* c_n - a^*)^*|| \\ &= ||R_{c_n}||\cdot||c_n a - a|| + ||a^* c_n - a^*||, \end{aligned}$$

and that the families $\{R_{c_n}\}_{n=1}^\infty$ and $\{L_{c_n}\}_{n=1}^\infty$ of right and left multiplication operators are uniformly bounded by the uniform boundedness theorem.) If M is a bound for $\{R_{c_n}\}_{n=1}^\infty$, then $||c_n||^2 = ||c_n^* c_n|| \leq ||c_n^*||\cdot||R_{c_n}|| \leq ||c_n^*||M = ||c_n||M$, so $\{c_n\}$ is bounded. Hence we can define $b = \Sigma_{n=1}^\infty c_n/2^n$. Then b is a strictly positive element of A (see (12.7)). By (12.9) A contains a countable increasing abelian a.i. $\{d_n\}$. Let A_o denote the maximal abelian *-subalgebra of A that contains $\{d_n\}$. Then $A_o = C_o(S)$, where S is the locally compact, σ-compact (12.2)), maximal ideal space of A_o. Hence A_o has a well-behaved countable a.i. $\{e_n\}$. We claim $\{e_n\}$ is an a.i. for A. Let

$a \in A$ and $\varepsilon > 0$. Choose a positive integer m so that $||a - d_m a|| < \varepsilon/2$ and then choose a positive integer N so that $||d_m - e_n d_m|| < \varepsilon/2(||a||+1)$ for all integers $n \geq N$. It follows that

$$||a - e_n a|| \leq ||(1 - e_n)(a - d_m a)|| + ||(d_m - e_n d_m)a|| < \varepsilon$$

for $n \geq N$. Hence $\{e_n\}$ is a well-behaved approximate identiy for A. □

We next characterize paracompactness in terms of well-behaved approximate identities.

(28.9) Theorem. Let X be a locally compact Hausdorff space. Then X is paracompact if and only if $C_o(X)$ has a well-behaved approximate identity bounded by one.

If X is paracompact it follows from (28.6) that $C_o(X)$ has a w.a.i.. The proof of the converse of (28.9) is nontrivial and relies on a sequence of lemmas.

Recall that a topological space X is zero-dimensional if the topology has a base of clopen sets. A map ψ from a topological space Y into X is perfect if it is a continuous closed surjection such that the inverse image of each point of X is compact. If Y is paracompact and $\psi : Y \to X$ is perfect, then it is well known that X is also paracompact (see Dugundji [74, p. 165]).

(28.10) Lemma. If Y is a zero-dimensional locally compact Hausdorff space, and $C_o(Y)$ has a w.a.i. $\{f_\lambda\}_{\lambda \in \Lambda}$, then there is a corresponding family $\{K_\lambda\}_{\lambda \in \Lambda}$ of compact-open subsets of Y such that:

(a) $\bigcup_{\lambda\in\Lambda} K_\lambda = Y$;

(b) if $\lambda_1 < \lambda_2$, then $K_{\lambda_1} \subset K_{\lambda_2}$;

(c) if $\lambda_o \in \Lambda$ and $\{\lambda_n\}$ is a strictly increasing sequence in Λ, then there is a positive integer n_o such that $K_{\lambda_o} \cap K_{\lambda_m} = K_{\lambda_o} \cap K_{\lambda_n}$ for $m, n \geq n_o$.

Proof. For each $\lambda \in \Lambda$, the set $U_\lambda = \{t \in X\colon f_\lambda(t) = 1\}$ is a compact subset of the open set $V_\lambda = \{t \in X\colon f_\lambda(t) > 1/2\}$. Choose a compact open set K_λ with $U_\lambda \subset K_\lambda \subset V_\lambda$. Then (a) holds, and, since $V_{\lambda_1} \subset U_{\lambda_2}$ for $\lambda_1 < \lambda_2$, (b) also holds. Since $\{f_\lambda\}_{\lambda\in\Lambda}$ is well-behaved, there is a positive integer N such that $f_{\lambda_o} f_{\lambda_n} = f_{\lambda_o} f_{\lambda_m}$ for $m, n \geq N$. Setting $n_o = N + 1$, we see that (c) holds. □

(28.11) Lemma. Let X and Y be locally compact Hausdorff spaces. If $C_o(X)$ has a w.a.i. and there is a perfect map of Y onto X, then $C_o(Y)$ also has a w.a.i..

Proof. The inverse image of a compact set under a perfect map is compact. Hence if $\{f_\lambda\}_{\lambda\in\Lambda}$ is a w.a.i. in $C_o(X)$, and $\psi : Y \to X$ is perfect, it follows that $\{f_\lambda \circ \psi\}_{\lambda\in\Lambda}$ is a w.a.i. for $C_o(Y)$ (see the Appendix.) □

(28.12) Definition. Let D be a set, Γ a directed set, and $\{U_\alpha\}_{\alpha\in\Gamma}$ a family of subsets of D. Then $\{U_\alpha\}_{\alpha\in\Gamma}$ is called a well-behaved cover of D if:

(a) $\bigcup_{\alpha\in\Gamma} U_\alpha = D$;

(b) if $\alpha_1 < \alpha_2$, then $U_{\alpha_1} \subset U_{\alpha_2}$;

(c) if $\alpha_0 \in \Gamma$ and $\{\alpha_n\}$ is a strictly increasing sequence in Γ, then there is a positive integer n_0 such that $U_{\alpha_0} \cap U_{\alpha_m} = U_{\alpha_0} \cap U_{\alpha_n}$ for $m, n \geq n_0$.

Remark. There is a simple method of constructing well-behaved covers of a set D. Indeed, let $\{V_\beta\}_{\beta\varepsilon B}$ be any decomposition of D into pairwise disjoint nonempty subsets. Let Γ be the collection of all finite subsets of B, directed by inclusion. For each $\alpha = \{\beta_1,\ldots,\beta_n\}$ in Γ, define $U_\alpha = \bigcup_{i=1}^{n} V_{\beta_i}$. Then (a), (b) and (c) of (28.12) hold for $\{U_\alpha\}_{\alpha\varepsilon\Gamma}$.

(28.13) Definition. Let D be a set. A well-behaved cover of D produced as in the preceding remark is called a decomposable cover of D.

(28.14) Definition. Two covers $\mathcal{C}$ and $\mathcal{D}$ of a set D are said to be equivalent if given $U \varepsilon \mathcal{C}$, there exists $W \varepsilon \mathcal{D}$ such that $U \subset W$; and given $W \varepsilon \mathcal{D}$, there exists $U \varepsilon \mathcal{C}$ such that $W \subset U$.

To see why we have introduced the preceding three definitions we now begin the proof of the converse of (28.9).

Proof of Theorem (28.9):

Suppose $C_0(X)$ has a w.a.i. $\{f_\lambda\}_{\lambda\varepsilon\Lambda}$. Let D denote the underlying set of X, endowed with the discrete topology, and let βZ denote the Stone-Čech compactification of a space Z. Then the identity map

$i : D \to X$ has a unique continuous extension $\psi : \beta D \to \beta X$ [74, p. 155]. Let $Y = \psi^{-1}(X)$, and let $\phi = \psi|Y$. Then we have:

(1) Y is a locally compact Hausdorff space, since Y is open in βD [74, p. 245].

(2) $D \subset Y \subset \beta D$; thus Y is extremally disconnected (if U is open in Y, then $\chi_{U \cap D}$ extends to a continuous F on βD; so $\overline{U} = \overline{U \cap D} = F^{-1}(0,\infty)$ is open.)

(3) ϕ is a perfect map of Y onto X, since ψ is perfect and ϕ is the restriction to a complete inverse image.

Applying (28.11) and (28.10) we obtain a family $\{K_\lambda\}_{\lambda\varepsilon\Lambda}$ of compact open subsets of Y satisfying (a), (b) and (c) of (28.10). For each $\lambda \varepsilon \Lambda$, let $H_\lambda = K_\lambda \cap D$; then $cl_{\beta D} H_\lambda = K_\lambda$. Let $\mathcal{L} = \{H_\lambda\}_{\lambda\varepsilon\Lambda}$. Then $\mathcal{L}$ is a well-behaved cover of D.

Now suppose (for the moment) that $\mathcal{L}$ is equivalent to some decomposable cover $\mathcal{D}$ arising from a decomposition $\{V_\beta\}_{\beta\varepsilon B}$ of D. Then

$$Y = \bigcup_{\lambda\varepsilon\Lambda} K_\lambda = \bigcup_{\lambda\varepsilon\Lambda} cl_{\beta D} H_\lambda$$

$$= \bigcup_{\alpha\varepsilon\Gamma} cl_{\beta D} U_\alpha = \bigcup_{\beta\varepsilon B} cl_{\beta D} V_\beta,$$

where the third equality holds because $\mathcal{L}$ and $\mathcal{D}$ are equivalent, and the last because each U_α is a finite union of sets V_β. But the sets $\{cl_{\beta D} V_\beta\}_{\beta\varepsilon B}$ are pairwise disjoint compact open subsets of βD, and this implies that Y is paracompact. Since $\phi : Y \to X$ is perfect, X will then also be paracompact.

Hence, the proof of (28.9) reduces to the following purely set-theoretic lemma of A. Hajnal:

III. MORE ABOUT APPROXIMATE IDENTITIES

(28.15) Lemma. A well-behaved cover of a set D is always equivalent to some decomposable cover of D.

Proof. Let $\mathcal{S}$ be a family of nonempty subsets of a set S which covers S. We shall say that $\mathcal{S}$ is a good cover of S if there is a function f which assigns to each finite collection $\{A_1,\ldots,A_n\}$ of distinct members of $\mathcal{S}$ a member $f(A_1,\ldots,A_n)$ of $\mathcal{S}$ in such a way that:

(i) $\bigcup_{i=1}^{n} A_i \subset f(A_1,\ldots,A_n)$;

(ii) $f(A_1,\ldots,A_{n-1}) \subset f(A_1,\ldots,A_{n-1},A_n)$; and

(iii) if $B \in \mathcal{S}$ and $\mathcal{S}' \subset \mathcal{S}$, there is a finite subcollection $\{A_1,\ldots,A_n\}$ of $\mathcal{S}'$ such that $B \cap (\bigcup\{W: W \in \mathcal{S}'\}) \subset f(A_1,\ldots,A_n)$.

Such a function f will be called a good function for $\mathcal{S}$.

Step 1. A well-behaved cover $\mathcal{L} = \{H_\lambda\}_{\lambda\in\Lambda}$ of a set D is a good cover of D.

Proof. Define a function g from the collection of finite subsets of Λ to Λ so that for any $\{\lambda_1,\ldots,\lambda_n\} \subset \Lambda$, $\lambda_i < g(\lambda_1,\ldots,\lambda_n)$ for all i and $g(\lambda_1,\ldots,\lambda_{n-1}) < g(\lambda_1,\ldots,\lambda_n)$. This is easily done by induction on n, the number of elements in the finite subset. We would like to define $f(H_{\lambda_1},\ldots,H_{\lambda_n})$ to be $H_{g(\lambda_1,\ldots,\lambda_n)}$, but there is a difficulty in that $H_\lambda = H_\mu$ for $\lambda \neq \mu$ might occur, leading to ambiguity in the definition. Proceed as follows: well-order Λ as $\{\lambda(\alpha)\}_{\alpha<\alpha_o}$ (this well-ordering of course has nothing to do with the partial order which Λ already has). If $P_1,\ldots,P_n$ are distinct members of $\mathcal{L}$, choose, for each i, the least α_i such that $H_{\lambda(\alpha_i)} = P_i$. Then define $f(P_1,\ldots,P_n)$ to be H_μ where $\mu = g(\lambda(\alpha_1),\ldots,\lambda(\alpha_n))$. It follows easily that (i) and (ii) hold. Suppose

(iii) fails for some $P_o \in \mathcal{L}$ and $\mathcal{L}' \subset \mathcal{L}$. By induction we can find a sequence $\{P_n\}$ in $\mathcal{L}'$ such that $P_o \cap P_n \not\subset f(P_1,\ldots,P_{n-1})$ for all n (P_1 is an arbitrary member of $\mathcal{L}'$). Let $P_n = H_{\lambda(\alpha_n)}$ as above. Then property (c) of a well-behaved cover is violated for the indices $\lambda(\alpha_o)$ and $g(\lambda(\alpha_1)) < g(\lambda(\alpha_1),\lambda(\alpha_2)) < \ldots < g(\lambda(\alpha_1),\ldots,\lambda(\alpha_n)) < \ldots$, a contradiction. Thus (iii) holds, and so f is a good function for $\mathcal{L}$.

Step 2. If $\mathcal{D}$ is a good cover of D, T is a nonempty subset of D, and $\mathcal{D}_T = \{A \cap T\colon A \in \mathcal{D}, A \cap T \neq \emptyset\}$, then $\mathcal{D}_T$ is a good cover of T.

Proof. Let $\{A_\alpha\}_{\alpha<\alpha_o}$ be a well-ordering of $\mathcal{D}$. If $B \in \mathcal{D}_T$, let $\alpha(B)$ be the least α such that $A_{\alpha(B)} \cap T = B$. If $B_1,\ldots,B_n \in \mathcal{D}_T$, define $h(B_1,\ldots,B_n) = f(A_{\alpha(B_1)},\ldots,A_{\alpha(B_n)}) \cap T$. It can be verified that h is a good function for $\mathcal{D}_T$ if f is a good function for $\mathcal{D}$.

Step 3. A good cover of a set is equivalent to a decomposable cover of that set.

Proof. We induct on the cardinality of a good cover $\mathcal{D}$. If $\text{card}\,\mathcal{D} \leq \aleph_o$, the assertion is easily established. Now suppose the result holds for good covers $\mathcal{D}$ of arbitrary sets, where $\text{card}\,\mathcal{D} < \kappa$ and $\kappa > \aleph_o$. Let $\text{card}\,\mathcal{D} = \kappa$, where $\mathcal{D}$ is a good cover of a set D, and let α_κ be the least ordinal of cardinal κ. Let f be a fixed good function for $\mathcal{D}$.

If $\mathcal{C} \subset \mathcal{D}$, we shall say that $\mathcal{C}$ is closed if $A_1,\ldots,A_n \in \mathcal{C}$ implies $f(A_1,\ldots,A_n) \in \mathcal{C}$. We shall construct a transfinite sequence $\{\mathcal{D}_\alpha\}_{\alpha<\alpha_\kappa}$ of closed subfamilies of $\mathcal{D}$ such that:

(1) $\alpha < \beta$ implies $\mathcal{D}_\alpha \subset \mathcal{D}_\beta$;

(2) $\mathcal{D}_\alpha = \bigcup_{\beta<\alpha} \mathcal{D}_\beta$ for limit ordinals α;

(3) $\bigcup_{\alpha<\alpha_\kappa} \mathcal{D}_\alpha = \mathcal{D}$; and

(4) card $\mathcal{D}_\alpha \leq$ card $\alpha + \aleph_0$ for all α.

Now $\mathcal{D}$ can be indexed as $\{A_\alpha\}$, where α runs over the set of nonlimit ordinals less than α_κ. If $\mathcal{S} \subset \mathcal{D}$, there is a smallest closed subfamily $\mathcal{C}(\mathcal{S})$ of $\mathcal{D}$ which contains $\mathcal{S}$, and it is not difficult to show that card $\mathcal{C}(\mathcal{S}) \leq \aleph_0 +$ card $\mathcal{S}$. Let $\mathcal{D}_1 = \mathcal{C}(\{A_1\})$. Suppose $\mathcal{D}_\alpha$ has been chosen for all $\alpha < \alpha_0$ so that $A_\alpha \in \mathcal{D}_\alpha$ for nonlimit ordinals α and (1), (2), (4) hold for $\alpha < \alpha_0$. If α_0 is a limit ordinal, let $\mathcal{D}_{\alpha_0} = \bigcup_{\alpha<\alpha_0} \mathcal{D}_\alpha$. If $\alpha_0 = \alpha_1 + 1$, let $\mathcal{D}_{\alpha_1} = \mathcal{C}(\mathcal{D}_{\alpha_0} \cup \{A_{\alpha_0}\})$. In this way the desired transfinite sequence is obtained.

We next let $S_\alpha = \bigcup\{U: U \in \mathcal{D}_\alpha\}$, and $Z_\alpha = S_{\alpha+1} \setminus S_\alpha$ for $\alpha < \alpha_\kappa$ (let $\mathcal{D}_0 = \emptyset$). Note that no member of $\mathcal{D}_\alpha$ meets Z_α. Let $\mathcal{M}_\alpha = \{B \cap Z_\alpha : B \in \mathcal{D}_{\alpha+1} \setminus \mathcal{D}_\alpha\}$. As in Step 2, one can show that $\mathcal{M}_\alpha$ is a good cover of Z_α for each nonempty Z_α. Since card $\mathcal{M}_\alpha \leq$ card $\mathcal{D}_{\alpha+1} < \kappa$, we have by induction that $\mathcal{M}_\alpha$ is equivalent to a decomposable cover $\mathcal{L}_\alpha$ of Z_α. Since D is the disjoint union of the sets Z_α, the family $\mathcal{L}$ of finite unions of all members of the collections $\mathcal{L}_\alpha$ is a decomposable cover of D. This proves part of Step 3. To complete the proof we consider:

<u>Step</u> 4. Finally, we show that $\mathcal{D}$ and $\mathcal{L}$ are equivalent.

<u>Proof</u>. If $V \in \mathcal{L}$, then $V = \bigcup_{i=1}^n V_{\alpha_i}$, where $V_{\alpha_i} \in \mathcal{L}_{\alpha_i}$. Then each $V_{\alpha_i} \subset U_{\alpha_i}$ for suitable $U_{\alpha_i} \in \mathcal{D}_{\alpha_i+1} \setminus \mathcal{D}_{\alpha_i}$, and $V \subset f(U_{\alpha_1}, \ldots, U_{\alpha_n}) \in \mathcal{D}$.

Conversely, we show by induction that if $W \in \mathcal{D}$, and α is the least ordinal such that $W \in \mathcal{D}_\alpha$, then $W \subset V$ for some $V \in \mathcal{L}$. For $\alpha = 1$ this is clear, since $\mathcal{D}_1$ and $\mathcal{L}_1$ are equivalent covers of Z_1. Suppose the result holds for all $\alpha < \alpha_o$. If α_o is a limit ordinal, then $\mathcal{D}_{\alpha_o} = \cup_{\alpha<\alpha_o} \mathcal{D}_\alpha$ and the result holds. Suppose $\alpha_o = \alpha_1 + 1$. Applying property (iii) of a good cover to $\mathcal{D}$, with $B = W$ and $\mathcal{S}' = \mathcal{D}_{\alpha_1}$, there exist $A_1,\ldots,A_n \in \mathcal{D}_{\alpha_1}$ such that $W \cap [\cup\{V\colon V \in \mathcal{D}_{\alpha_1}\}] = W \cap S_{\alpha_1} \subset f(A_1,\ldots,A_n)$. Since $\mathcal{D}_{\alpha_1}$ is a closed family, $f(A_1,\ldots,A_n) \in \mathcal{D}_{\alpha_1}$. Thus by induction, $W \cap S_{\alpha_1} \subset V_1$ for some $V_1 \in \mathcal{L}$. Since $W \in \mathcal{D}_{\alpha_1+1} \setminus \mathcal{D}_{\alpha_1}$, we have $W \cap (S_{\alpha_1+1} \setminus S_{\alpha_1}) = W \cap Z_{\alpha_1} \in \mathcal{M}_{\alpha_1}$. Since $\mathcal{M}_{\alpha_1}$ and $\mathcal{L}_{\alpha_1}$ are equivalent covers of Z_{α_1}, there is a member V_2 of $\mathcal{L}$ with $W \cap (S_{\alpha_1+1} \setminus S_{\alpha_1}) \subset V_2$. Then $U \subset V_1 \cup V_2$, which is in $\mathcal{L}$ because a decomposable cover is closed under finite unions. This completes the proof of the lemma and also the proof of (28.9). □

If Ω_o denotes the space of ordinals less than the first uncountable ordinal with the order topology, then Ω_o is a locally compact, nonparacompact, Hausdorff space. It follows from (28.9) that $C_o(\Omega_o)$ does not have a well-behaved approximate identity.

(28.16) Proposition. Let X and Y be locally compact Hausdorff spaces. The algebra $C_o(X \times Y)$ has a w.a.i. if and only if each of the algebras $C_o(X)$ and $C_o(Y)$ has a w.a.i..

Proof. If $\{e_\lambda\}_{\lambda\varepsilon\Lambda}$ is a w.a.i. for $C_o(X \times Y)$ and $t_o \varepsilon Y$, then the net of functions $\{f_\lambda\}_{\lambda\varepsilon\Lambda}$ defined by $f_\lambda(s) = e_\lambda(s,t_o)$ is a w.a.i. for $C_o(X)$. Similarly one obtains a w.a.i. for $C_o(Y)$.

On the other hand, suppose $\{e_\lambda\}_{\lambda\varepsilon\Lambda}$ is a w.a.i. for $C_o(X)$ and $\{f_\alpha\}_{\alpha\varepsilon\Gamma}$ is a w.a.i. for $C_o(Y)$. If $f \varepsilon C_o(X)$ and $g \varepsilon C_o(Y)$, let $f \otimes g$ be the function on $X \times Y$ defined by $(f \otimes g)(s,t) = f(s)g(t)$. Then $f \otimes g \varepsilon C_o(X \times Y)$. Because the algebra generated by the set $\{f \otimes g: f \varepsilon C_o(X), g \varepsilon C_o(Y)\}$ is dense in $C_o(X \times Y)$ by the Stone-Weierstrass theorem, the net $\{e_\lambda \otimes f_\alpha\}$ with directed set all pairs (λ,α), where $(\lambda,\alpha) > (\lambda',\alpha')$ if $\lambda > \lambda'$ and $\alpha > \alpha'$, is an approximate identity for $C_o(X \times Y)$ which is easily seen to be well-behaved (see Appendix.) □

29. Quasicentral approximate identities.

In this section we shall study the useful notion of quasicentral approximate identity. Such approximate identities have recently played a key role in unifying and simplfying several important areas of C^*-algebra and operator theory (see Akemann and Pedersen [5] and Arveson [16]).

Let A be a Banach algebra with bounded approximate identity bounded by one. A double centralizer on A is a pair (R,S) of functions from A into A such that

$$aR(b) = S(a)b \qquad (a, b \varepsilon A).$$

Let M(A) denote the set of all double centralizers of A. If (R,S) is in M(A), then R and S are bounded linear operators on A and $||R|| = ||S||$. Hence M(A) under the usual linear operations, multiplication

$(R_1,S_1)(R_2,S_2) = (R_1R_2,S_1S_2)$, and norm $||(R,S)|| = ||R||$, is a Banach algebra with identity, called the <u>double-centralizer algebra of</u> A.

If we define a map $\mu : A \to M(A)$ by $\mu(a) = (L_a,R_a)$, where $L_a(b) = ab$ and $R_a(b) = ba$ for all $b \in A$, then μ is an isometric isomorphism of A into M(A) and $\mu(A)$ is a closed two-sided ideal of M(A). Hence A will usually be viewed as a closed ideal of M(A).

When A is a C^*-algebra, M(A) becomes a C^*-algebra under the involution $(R,S)^* = (S^*,R^*)$, where $R^*(a) = (R(a^*))^*$ and $S^*(a) = (S(a^*))^*$ for all $a \in A$, and μ is an isometric *-isomorphism of A into M(A). For a more detailed account of M(A) we refer the reader to Johnson [135] and Busby [46].

(29.1) <u>Definition</u>. A Banach algebra A is said to have a <u>bounded quasicentral approximate identity</u> bounded by one if there is a net $\{e_\lambda\}_{\lambda\in\Lambda}$ in A such that $||e_\lambda|| \leq 1$ for all $\lambda \in \Lambda$ and $||e_\lambda a - a||$, $||ae_\lambda - a||$, and $||e_\lambda b - be_\lambda||$ tend to zero as λ runs through Λ for all $a \in A$ and all $b \in M(A)$.

If a Banach algebra A has a bounded quasicentral approximate identity $\{e_\lambda\}_{\lambda\in\Lambda}$, and B is any Banach algebra containing A as a two-sided ideal, then

$$||e_\lambda b - be_\lambda|| \to 0 \quad (b \in B).$$

Indeed, let $b_o \in B$. If $L_b(a) = ba$ and $R_b(a) = ab$ for all $a \in A$ and $b \in B$, then the map $\psi : B \to M(A)$ defined by $\psi(b) = (L_b,R_b)$ is a norm reducing homomorphism that is isometric on A. Thus $||e_\lambda b_o - b_o e_\lambda|| = ||\psi(e_\lambda)\psi(b_o) - \psi(b_o)\psi(e_\lambda)|| = ||e_\lambda\psi(b_o) - \psi(b_o)e_\lambda|| \to 0$ as λ runs through Λ. Hence our definition of quasicentral approximate identity includes the original definition as given by Arveson [16] and Akemann and Pedersen [5].

As usual, we can translate the net definition of bounded quasicentral approximate identity into the language of ε, finite subsets of A, and finite subsets of $M(A)$ (see (1.2)).

Let A be a Banach algebra with dual space A^*. The <u>notation</u> $A^* = A\cdot A^*$ (resp. $A^* = A^*\cdot A$) means that if $f \in A^*$ then there exist $a \in A$ and $g \in A^*$ such that $f = a\cdot g$ (resp. $f = g\cdot a$), where $a\cdot g(x) = g(xa)$ and $g\cdot a(x) = g(ax)$ for all $x \in A$. In particular, if $A^* = A^*\cdot A + A\cdot A^*$ then for $f \in A^*$ there exist $a, b \in A$ and $g, h \in A^*$ such that $f(x) = g(ax) + h(xb)$ for all $x \in A$.

(29.2) <u>Theorem</u>. Let A be a Banach algebra with bounded approximate identity $\{e_\lambda\}_{\lambda\in\Lambda}$ bounded by one. If $A^* = A^*\cdot A + A\cdot A^*$, then A has a bounded quasicentral approximate identity bounded by one which is contained in the convex hull of $\{e_\lambda\}_{\lambda\in\Lambda}$.

<u>Proof</u>. Let E be the convex hull of $\{e_\lambda\}_{\lambda\in\Lambda}$, and for each $\lambda_1 \in \Lambda$ let $E(\lambda_1)$ be the convex hull of the set $\{e_\lambda: \lambda \in \Lambda, \lambda \geq \lambda_1\}$. We shall show that there is a bounded quasicentral approximate identity for A in E by showing that for each $\varepsilon > 0$, each positive integer n, each subset $\{a_1,\ldots,a_n\}$ of A, and each subset $\{b_1,\ldots,b_n\}$ of $M(A)$, there is an $e \in E$ with $||ea_j - a_j|| < \varepsilon$, $||a_je - a_j|| < \varepsilon$, and $||eb_j - b_je|| < \varepsilon$ for $j = 1,\ldots,n$.

Let $\varepsilon > 0$, let $a_1,\ldots,a_n \in A$, let $b_1,\ldots,b_n \in M(A)$, and choose $\lambda_o \in \Lambda$ such that $||e_\lambda a_j - a_j|| < \varepsilon$ and $||a_je_\lambda - a_j|| < \varepsilon$ for $\lambda \geq \lambda_o$.

§29. QUASICENTRAL APPROXIMATE IDENTITIES

Let W be the closure of $\{\bigoplus_{j=1}^{n}(eb_j - b_je) : e \in E(\lambda_o)\}$ in the Banach space direct sum of n-copies of A. Then W is a closed convex set because $E(\lambda_o)$ is convex and the map $e \to \bigoplus_{j=1}^{n}(eb_j - b_je)$ from A into the direct sum of n-copies of A is linear. It follows from the Hahn-Banach (separation) theorem that if $0 \notin W$, then there is an $(f_1,\ldots,f_n) \in (A^n)^* = (A^*)^n$ such that

$$\left|\sum_{j=1}^{n} f_j(eb_j - b_je)\right| \geq 1 \tag{1}$$

for all $e \in E(\lambda_o)$. Now, by hypothesis, for each j there are functionals $g_j, h_j \in A^*$ and elements $c_j, d_j \in A$ such that

$$f_j(x) = g_j(c_jx) + h_j(xd_j)$$

for all $x \in A$. Now

$$f_j(e_\lambda b_j - b_je_\lambda) = g_j(c_je_\lambda b_j - c_jb_je_\lambda) + h_j(e_\lambda b_jd_j - b_je_\lambda d_j)$$

which converges to zero as λ runs through Λ with $\lambda \geq \lambda_o$, because e_λ is an approximate identity for A. This contradicts inequality (1), and shows that $0 \in W$, which completes the proof. □

(29.3) Theorem. Every C^*-algebra A has a bounded quasicentral approximate identity bounded by one.

Proof. In view of (29.2) it suffices to show that $A^* = A^*\cdot A = A\cdot A^*$. Let f be a positive linear functional in A^*. By Rickart [222, (4.5.14), p. 219] f is representable; that is, there exists a Hilbert space H, a continuous *-representation $a \to T_a$ of A on H, and a topologically cyclic vector $\xi \in H$ such that $f(a) = (T_a \xi | \xi)$ for all $a \in A$. Let $\{e_\lambda\}_{\lambda \in \Lambda}$ be an approximate identity for A (12.4). Since $\xi = \lim_{n\to\infty} T_{a_n} \xi$ for some sequence $\{a_n\}$ in A, it follows that $T_{e_\lambda} \xi = \xi$. Since H is a Banach A-module we have by the fundamental factorization theorem (16.1) that $\xi = T_a \xi_0$ for some $a \in A$ and $\xi_0 \in H$. Define g on A by the formula $g(b) = (T_b \xi_0 | \xi_0)$ for each $b \in A$ and note that $g \in A^*$ and $f = a \cdot g \cdot a^*$.

Now assume that f is <u>any</u> element of A^*. Using the standard fact that f can be expressed as a linear combination of four positive linear functionals on A (see Sakai [239, (1.17.1), p. 42]), we see that $\lim_\lambda e_\lambda \cdot f = \lim_\lambda f \cdot e_\lambda = f$. Hence, again by theorem (16.1), there exist elements $a, b \in A$ and linear functionals $g_1, g_2 \in A^*$ such that $f = a \cdot g_1 = g_2 \cdot b$ and the proof is complete. □

(29.4) <u>Corollary</u>. A separable C^*-algebra contains a sequential, increasing, abelian, well-behaved, bounded quasicentral approximate identity bounded by one.

<u>Proof</u>. See (12.4), (12.9), (28.24) and (29.3). □

The next result depends on Theorem (21.1) and its proof.

(29.5) <u>Theorem</u>. Let A be a closed ideal in a Banach algebra B, and let A have a bounded countable approximate identity. If the quotient B/A is separable, and if A has a bounded quasicentral approximate identity bounded by one, then there is a norm continuous semigroup $t \to a^t$ from

$(0,\infty)$ into A such that $||a^t|| \leq 1$ for all $t > 0$, and $||a^t x - x||$, $||xa^t - x||$, and $||a^t b - ba^t||$ tend to zero as t decreases to zero for all $x \in A$ and $b \in B$.

<u>Proof</u>. Let $\{c_1, c_2, \ldots\}$ be a countable subset of B such that the set $\{c_1 + A, c_2 + A, \ldots\}$ is dense in B/A and let $\{x_n\}$ be a bounded countable approximate identity for A. By (17.5) there exist elements x and y in A and a sequence $\{z_n\}$ in A such that $x_j = xz_j y$ for $j = 1,2,\ldots$. We shall apply Theorem (21.1) to x in the left Banach A-module A and to y in the right Banach A-module A at the same time placing a further restriction on the choice of the sequence $\{e_n\}$ used in the proof of (21.1).

Let Y be the Banach space of all sequences from B that converge to zero, and regard Y as a Banach A-module in the natural way. Let u be the element $\{c_j ||c_j||^{-1} j^{-1}\}$ in Y. By hypothesis we can choose elements $e \in A$ such that $||e|| \leq 1$ and $||ec_j - c_j e||$ are arbitrarily small for a finite number of j. Hence we can choose $e \in A$ such that $||e|| \leq 1$ and $||eu - ue||$ is arbitrarily small. In the proof of (21.1) we may ensure that $||e_{n+1}u - ue_{n+1}||$ is so small that

$$2C_1||u||\cdot||(1-e)f|| + 2C_2||u||\cdot||f(1-e)|| + C_4||eu-ue|| \leq 2^{-n-1}. \quad (4')$$

Using Theorem (21.1), we now define $a^t = \theta(w^t)$ for $t > 0$, where $w(z) = z$ for all $z \in \Delta$. The continuity of $t \to a^t$ for $t > 0$ follows because $w^t \in F_\alpha$ if $t > \alpha$ by (21.1)(iv). By (21.1)(vii) and the remarks above $\{a^t: t > 0, t \to 0\}$ is a bounded approximate identity for A bounded by one.

MORE ABOUT APPROXIMATE IDENTITIES

To show that $||a^t c - ca^t|| \to 0$ as $t \to 0$ $(t > 0)$ for all $c \in B$ it is sufficient to show that $||a^t c_j - c_j a^t|| \to 0$ because $\{c_j + A\}_{j=1}^{\infty}$ is dense in B. This will follow if we show that $||a^t u - ua^t|| \to 0$ as $t \to 0$ from the definition of u. We shall deduce this from (4') and the other inequalities in the proof of (21.1) in a similar way to that in which (21.1)(vii) was proved.

By (4') and Lemma (21.2)(iii) for $g \in F$

$$||G(\mu_{n+1}-f_{n+1})\cdot u-u\cdot G(\mu_{n+1}-f_{n+1})-G(\mu_n-f_n)\cdot u+u\cdot G(\mu_n-f_n)|| \leq ||G||_{\Gamma(n)}\cdot 2^{-n-1}$$

$$\text{for } n = 1,2,3,\dots .$$

Using this inequality we obtain

$$||\theta(g)\cdot u-u\cdot\theta(g)|| \leq 2^{-n}||G||_{\Gamma(n)}+||G(\mu_n-f_n)\cdot u-u\cdot G(\mu_n-f_n)||,$$

where $\Gamma(n)$ is the circle with center μ_n and radius r used in inequality (6) of the proof of (21.1). Using the inequality

$$||G(\mu_n-f_n)\cdot u-u\cdot G(\mu_n-f_n)|| \leq 2\cdot||u||\cdot||G(\mu_n-f_n)-1||$$

and the estimate on $||G(\mu_n-f_n)-1||$ obtained in proving inequality (10) of the proof of (21.1), we have

$$||\theta(g)\cdot u-u\cdot\theta(g)|| \leq 2^{-n}||g||_{\Delta}+4||u||\cdot K^n\sup\{|1-g(v)|: v \in \Delta,\ |v| \geq (2K^n+1)^{-1}\}$$

because $||G||_{\Gamma(n)} \leq ||G||_H = ||g||_{\Delta}$ by the maximum modulus principle. Hence

$$||a^t u - ua^t|| \leq 2^{-n}+4||u||\cdot K^n\cdot\sup\{|1-v^t|: v \in \Delta,\ |v| \geq (2K^n+1)^{-1}\}$$

for $n = 1,2,3,\dots$ and $t > 0$. Choosing n so large that 2^{-n} is small and then choosing t very small and positive so the second term in the above inequality is small, we see that $||a^t u - ua^t|| \to 0$ as $t \to 0$. □

A result similar to (29.5) is the following whose proof we simply sketch.

(29.6) Theorem. Let A be a Banach algebra. Then there is a continuous semigroup $t \to a^t$ from $(0,\infty)$ into A such that $||a^t|| \leq 1$ and $\sigma(a^t) \subset \{z \in C: |\arg z| \leq t\}$ for all $t > 0$, and $||a^t x - x|| + ||xa^t - x|| \to 0$ as $t \to 0$ if and only if A has a bounded countable approximate identity bounded by one.

Proof. We assume A has a bounded countable approximate identity as the converse is straightforward. Using techniques like those in the proof of (29.5), with $a^t = \theta(w^{2t/\pi})$ for all $t > 0$, the results follow as in (29.5) except for the spectral condition, which follows from (21.1)(ii). □

30. Compact operators.

For any Banach space X we denote by $B'(X)$ the algebra of compact operators on X, i.e., bounded linear maps $T : X \to X$ such that T maps the unit ball of X (and hence any bounded subset of X) into a relatively compact subset of X. It is well known that $B'(X)$ is a closed two-sided ideal in the algebra $B(X)$ of all bounded linear maps on X. Furthermore, $B'(X)$ has an identity if and only if X is finite-dimensional.

If X is a Hilbert space, then $B'(X)$ is a closed two-sided *-ideal in $B(X)$. Hence, in this case, $B'(X)$ is a C^*-algebra and therefore has a two-sided approximate identity bounded by one. It is not known whether $B'(X)$, for a general Banach space X, always has a bounded approximate identity. However we can prove this in certain cases and the results below cover most of the classical Banach spaces.

A sequence $\{e_n\}$ in a Banach space X is called a <u>basis</u> of X if for every $x \in X$ there is a unique sequence $\{\alpha_n\}$ of scalars such that

$$x = \sum_{n=1}^{\infty} \alpha_n e_n.$$

If $\{e_n\}$ is a basis for X we define the <u>coefficient functionals</u> e_n^* on X by $e_n^*(x) = \alpha_n$ if $x = \sum_{n=1}^{\infty} \alpha_n e_n$. S. Banach showed that the coefficient functionals of a basis for a Banach space are necessarily continuous. Hence $e_n^* \in X^*$ and we will write

$$x = \sum_{n=1}^{\infty} \alpha_n e_n = \sum_{n=1}^{\infty} e_n^*(x) e_n = \sum_{n=1}^{\infty} (x, e_n^*) e_n.$$

The <u>projections</u> P_i, $i = 1,2,\ldots$, <u>associated with a basis</u> $\{e_n\}$ are defined by

$$P_i\left(\sum_{n=1}^{\infty} \alpha_n e_n\right) = \sum_{n=1}^{i} \alpha_n e_n.$$

Obviously, the P_i are bounded linear operators on X of finite rank. Since $x = \lim_i P_i x$ we have

$$||x|| \leq \sup_i ||P_i x|| < \infty \quad \text{for all } x \in X.$$

Hence by the uniform boundedness theorem,

$$\sup_i ||P_i|| < \infty.$$

The number $\sup_i ||P_i||$ is called the <u>basis constant</u> of $\{e_n\}$.

A Banach space X is said to have the <u>approximation property</u> if for every compact set $K \subset X$ and every $\varepsilon > 0$ there exists a finite rank operator F on X such that

$$||x - Fx|| < \varepsilon \quad \text{for all} \quad x \in K.$$

If, moreover, there exists a constant $\lambda > 0$ which is independent of K and ε such that F can be chosen with $||F|| \leq \lambda$ then X is said to have the <u>bounded approximation property</u> with bound λ.

It is easy to see that the existence of a basis $\{e_n\}$ in a Banach space X implies that X has the bounded approximation property. Indeed, let $\{P_i\}$ be the projections associated with the basis $\{e_n\}$ and set $\lambda = \sup_i ||P_i||$. Then for every compact set $K \subset X$ and every $\varepsilon > 0$ we can find $i_o = i_o(\varepsilon,K)$ such that the distance between x and the linear span of $\{e_1,\ldots,e_{i_o}\}$ is less than $\varepsilon/2\lambda$ for all $x \in K$. Then

$$||x - P_i x|| < \varepsilon \quad \text{for all} \quad i \geq i_o \quad \text{and all} \quad x \in K.$$

(30.1) <u>Proposition</u>. Let X be a Banach space. If X has the approximation property then the Banach algebra $B'(X)$ of compact operators on X has a left approximate identity. If X has the bounded approximation property with bound λ then the Banach algebra $B'(X)$ has a left approximate identity bounded by λ.

<u>Proof</u>. Let $T_1,\ldots,T_n$ be a finite number of compact operators on X. Then the closure of

$$\{T_i x \colon x \in X, \ ||x|| \leq 1, \ i = 1,\ldots,n\}$$

is a compact subset of X. Hence, by the approximation property of X, for every $\varepsilon > 0$ there exists a finite rank operator $F = F(\varepsilon, T_1, \ldots, T_n)$ on X such that

$$||T_i - FT_i|| = \sup_{\substack{x \in X \\ ||x|| \leq 1}} ||T_i x - FT_i x|| < \varepsilon, \quad i = 1,2,\ldots,n.$$

Thus the Banach algebra $B'(X)$ has a left approximate identity. If X has the bounded approximation property with bound λ then the F's may be chosen with $||F|| \leq \lambda$; i.e., the Banach algebra $B'(X)$ has a left approximate identity bounded by λ. □

(30.2) Theorem. Let X be a Banach space with a basis $\{e_n\}$. Then the associated sequence $\{P_i\}$ of finite rank projections is a bounded left approximate identity in the Banach algebra $B'(X)$ of compact operators on X.

Proof. If T is a compact operator on X then

$$||T - P_i T|| = \sup_{\substack{x \in X \\ ||x|| \leq 1}} ||Tx - P_i Tx|| \to 0 \quad \text{as} \quad i \to \infty. \;\square$$

In 1973 P. Enflo constructed a separable Banach space without the approximation property and thus without a basis. Using Enflo's example, T. Figiel and W. B. Johnson gave an example of a Banach space which has the approximation property but fails to have the bounded approximation property and thus also has no basis. It is still open whether for separable Banach spaces the existence of a basis is equivalent to the bounded approximation property.

§30. COMPACT OPERATORS

It is interesting to note that A. Grothendieck proved that if X is reflexive or X is a separable dual space then the approximation property for X implies the bounded approximation property with $\lambda = 1$.

(30.3) Remark. Let X be a reflexive or separable dual Banach space. If X has the approximation property then the Banach algebra $B'(X)$ of compact operators on X has a left approximate identity bounded by one.

To study the existence of right approximate identities let us recall the definition of a shrinking basis. A basis $\{e_n\}$ for a Banach space X is shrinking iff the associated sequence $\{e_n^*\}$ of coefficient functionals is a basis for the dual space X^*. In 1971 W. B. Johnson, H. P. Rosenthal, and M. Zippin showed that a Banach space X has a shrinking basis if X^* has a basis.

(30.4) Theorem. Let X be a Banach space with a shrinking basis $\{e_n\}$. Then the associated sequence $\{P_i\}$ of finite rank projections is a bounded two-sided approximate identity in the Banach algebra $B'(X)$ of compact operators on X.

Proof. We know that $\{P_i\}$ is a bounded left approximate identity in $B'(X)$. Since $\{e_n\}$ is a shrinking basis of X, the dual projections

$$P_i^*: X^* \to X^*$$

$$\sum_{n=1}^{\infty} \beta_n e_n^* \to \sum_{n=1}^{i} \beta_n e_n^*$$

are just the projections associated with the dual basis $\{e_n^*\}$ of X. Hence $\{P_i^*\}$ is a left approximate identity in $B'(X^*)$ and so

$$||T - TP_i|| = ||T^* - P_i^*T^*|| \to 0 \quad \text{as} \quad i \to \infty$$

for every $T \in B'(X)$. Here we have made use of the fact that the dual operator of a compact operator is also compact. Thus $\{P_i\}$ is a bounded two-sided approximate identity in $B'(X)$. □

(30.5) Example. Let p be a real number, $1 < p < \infty$. Let ℓ^p be the Banach space of all sequences $x = (\xi_n)_{n\in N}$ of complex numbers for which the infinite series $\sum_{n=1}^{\infty} |\xi_n|^p$ converges with norm defined by

$$||x||_p = (\sum_{n=1}^{\infty} |\xi_n|^p)^{1/p}.$$

Let the conjugate exponent q be defined by $\frac{1}{p} + \frac{1}{q} = 1$. Then ℓ^q is the dual space of ℓ^p. Obviously, the natural basis $\{e_n\}$ of ℓ^p, where

$$e_n = (0,\ldots,0,1,0,\ldots), \quad n = 1,2,\ldots,$$

is shrinking. Hence the associated sequence $\{P_i\}$ of finite rank projections

$$P_i : \ell^p \to \ell^p$$

$$(\xi_1,\xi_2,\ldots) \to (\xi_1,\ldots,\xi_i,0,0,\ldots)$$

is a two-sided approximate identity in the Banach algebra $B'(\ell^p)$ of compact operators on ℓ^p. Since $||P_i|| = 1$, $i = 1,2,\ldots$, the approximate identity is bounded by one. □

(30.6) Example. The Banach algebra $B'(C(T))$ of compact operators on the Banach space $C(T)$ of continuous functions on the circle group T of complex numbers of absolute value one has a two-sided approximate identity bounded by one.

Proof. For each n we can write T as the union of n disjoint congruent half open arcs $T_1,\ldots,T_n$. Choose nonnegative continuous functions $g_1,\ldots,g_n$ on T such that at each point at most two of the g_i are nonzero, the support of each g_i lies in T_i and the two adjacent arcs, and $\sum_i g_i(t) = 1$ for all $t \in T$. For each measure $\mu > \lambda$, where λ is Lebesgue measure, define

$$E_{n,\mu} f = \sum_i (fg_i,\mu)(g_i,\mu)^{-1} g_i, \quad f \in C(T).$$

A simple uniform continuity argument shows that as $n \to \infty$, $E_{n,\mu} f \to f$ uniformly with respect to μ. Thus ordering the measures μ in the usual way we see that the directed net $\{E_{n,\mu}\}$ converges to the identity pointwise on $C(T)$ and $||E_{n,\mu}|| \leq 1$ so that $\{E_{n,\mu}\}$ converges to the identity uniformly on compact sets in $C(T)$. Hence $\{E_{n,\mu}\}$ is a left approximate identity in $B'(C(T))$ bounded by one. In particular, the finite rank operators are dense in $B'(C(T))$.

Let $\nu_1,\ldots,\nu_p \in M(T)$, suppose $\mu > \lambda$ and $\mu > |\nu_j|$, $j = 1,\ldots,p$, so that $\nu_j = \phi_j \mu$ for $\phi_j \in L^1(\mu)$. We then have

$$E^*_{n,\mu} \nu_j = (\sum_i (g_i,\phi_j\mu)(g_i,\mu)^{-1} g_i)\mu.$$

Thus $\{E^*_{n,\mu}\}$ is a bounded sequence of operators in $M(T)$ for which $L^1(\mu)$ is an invariant subspace and such that $E^*_{n,\mu}(\phi\mu) \to \phi\mu$ for all ϕ in $C(T)$ and hence for all ϕ in $L^1(\mu)$ because $C(T)$ is dense in $L^1(\mu)$. This shows that $E^*_{n,\mu}\nu_j \to \nu_j$ as $n \to \infty$ for all j so that some subnet $\{E^*_\alpha\}$ of $\{E^*_{n,\mu}\}$ has $E^*_\alpha \nu \to \nu$ for all $\nu \in M(T)$. If F is the finite rank operator

$$Ff = \sum_j (f,\mu_j)f_j,$$

where $\mu_j \in M(T)$, $f_j \in C(T)$, $j = 1,\ldots,q$, then

$$FE_\alpha f = \sum_j (f,E^*_\alpha\mu_j)f_j$$

so that $FE_\alpha \to F$. Since $||E_\alpha|| \leq 1$ and the finite rank operators are dense in $B'(C(T))$ this shows that $\{E_\alpha\}$ is a right approximate identity bounded by one. Since $\{E_\alpha\}$ is a subnet of a left approximate identity it is a two-sided approximate identity. □

31. Abstract Segal algebras.

In this section we shall study approximate identities in Banach algebras which are ideals in another Banach algebra.

(31.1) Definition. Let $(A,||\cdot||_A)$ be a Banach algebra. A Banach algebra $(B,||\cdot||_B)$ is called an abstract Segal algebra in $(A,||\cdot||_A)$ if B is a dense left ideal in $(A,||\cdot||_A)$ and if for some constant $C > 0$,

$$||b||_A \leq C||b||_B \quad \text{for all } b \in B.$$

If the algebra A is semisimple the last norm-condition is fulfilled automatically.

(31.2) Proposition. Let B be a dense left ideal in a semisimple Banach algebra A. Then there exists a constant $C > 0$ such that

$$||b||_A \leq C||b||_B \quad \text{for all} \quad b \in B.$$

Proof. We prove that the embedding $(B,||\cdot||_B) \hookrightarrow (A,||\cdot||_A)$ is a closed, and hence continuous, map. Assume that $\{b_n\} \subset B$, $a \in A$, $||b_n||_B \to 0$, and $||b_n - a||_A \to 0$. We have to show that $a = 0$.

Let I be any maximal modular left ideal of A. Then there is a $u \in A$ such that $A(1 - u) \subset I$. Consider the left regular representation π of A on the quotient space A/I. Fix any $\xi \in A/I$, $\xi \neq 0$. Then A acts strictly irreducibly on A/I and $\pi(A)\xi = A/I$. By assumption, B is dense in A. Hence, by the continuity of the left regular representation π,

$$\overline{\pi(B)\xi}^{A/I} = \pi(A)\xi = A/I.$$

This implies that $\pi(B)\xi \neq \{0\}$. Since $\pi(B)\xi$ is $\pi(A)$-invariant, we have $\pi(B)\xi = A/I$. Thus B acts strictly irreducibly on A/I.

Let P be the kernel of this representation of B on A/I. P is a primitive ideal of B, and therefore P is closed in B. Hence the induced representation of the quotient algebra B/P on A/I is a faithful strictly irreducible representation of B/P into the bounded operators on A/I. A theorem of B. E. Johnson asserts that such a representation is continuous.

Since $||b_n + P||_B \leq ||b_n||_B \to 0$, then $||b_nu + I||_A =$

$||(b_n + P)(u + I)||_A \to 0$. Also $||(b_n - a)u + I||_A \leq ||(b_n - a)u||_A \leq ||b_n - a||_A||u||_A \to 0$. It follows that $au + I = 0$, and thus $a = au + (a - au) \in I$. Then a must be in every modular maximal left ideal of A, so that by the semisimplicity of A, $a = 0$. □

The next result implies that an abstract Segal algebra B in a Banach algebra A is a left Banach A-module.

(31.3) Proposition. Let B be a left ideal in a Banach algebra A. Assume that there exists a constant $C > 0$ such that $||b||_A \leq C||b||_B$ for all $b \in B$. Then there exists a constant $M > 0$ such that

$$||ab||_B \leq M||a||_A||b||_B \quad \text{for all} \quad a \in A,\ b \in B.$$

Proof. We remark here that the norm $||\cdot||_B$ need not be submultiplicative, i.e., it is simply a Banach space norm. Now for $b \in B$ consider the mapping

$$R_b\colon (A, ||\cdot||_A) \to (B, ||\cdot||_B)$$
$$a \to ab.$$

We prove that R_b is continuous by showing that R_b is a closed map. Assume that $\{a_n\} \subset A$, $c \in B$, $||a_n||_A \to 0$, and $||R_b(a_n) - c||_B \to 0$. Then $||a_nb - c||_A \leq C||a_nb - c||_B \to 0$. Also $||a_nb||_A \to 0$, and therefore $c = 0$.

Now since R_b is continuous, for each $b \in B$ there exists a constant M_b such that

$$||ab||_B \leq M_b||a||_A.$$

For $a \in A$ consider the operator map

$$L_a: (B, ||\cdot||_B) \to (B, ||\cdot||_B)$$

$$b \to ab$$

We prove that L_a is continuous by showing that L_a is a closed map. Assume that $\{b_n\} \subset B$, $c \in B$, $||b_n||_B \to 0$, and $||L_a(b_n) - c||_B \to 0$. Then $||ab_n - c||_A \leq C||ab_n - c||_B \to 0$. Also $||ab_n||_A \leq C||ab_n||_B \to 0$, and therefore $c = 0$. Thus L_a is continuous. Set

$$|L_a| = \sup\{||L_a(b)||_B : b \in B, ||b||_B \leq 1\}.$$

Let

$$\Omega = \{L_a : a \in A, ||a||_A \leq 1\}.$$

Then

$$||L_a(b)||_A \leq C||L_a(b)||_B = C||ab||_B \leq CM_b||a||_A \leq CM_b$$

for all $b \in B$ and $L_a \in \Omega$. Hence by the uniform boundedness theorem there exists a constant $M > 0$ such that

$$|L_a| \leq M \text{ for all } L_a \in \Omega.$$

Thus for all $a \in A$ with $||a||_A \leq 1$ and all $b \in B$

$$||ab||_B = ||L_a(b)||_B \leq M||b||_B$$

and so $||ab||_B \leq M||a||_A||b||_B$ for all $a \in A$, $b \in B$. □

As a converse to the preceding results we prove the following:

(31.4) Proposition. Let B be a subalgebra of a Banach algebra A.

III. MORE ABOUT APPROXIMATE IDENTITIES

Assume that

i) there exists $C > 0$ such that $||b||_A \leq C||b||_B$ for all $b \in B$, and

ii) there exists $M > 0$ such that $||ab||_B \leq M||a||_A||b||_B$ for all $a, b \in B$.

Then B is a left ideal in $\overline{B}^A$ and

$$||ab||_B \leq M||a||_A||b||_B \quad \text{for all} \quad a \in \overline{B}^A,\ b \in B.$$

<u>Proof</u>. Let $a \in \overline{B}^A$ and $b \in B$. Choose $\{a_n\} \subset B$ such that $||a_n - a||_A \to 0$. Then $||a_nb - a_mb||_B \leq M||a_n - a_m||_A||b||_B \to 0$, so that $\{a_nb\}$ is Cauchy in B. Hence there exists $c \in B$ such that $||a_nb - c||_B \to 0$. Then $||a_nb - c||_A \leq C||a_nb - c||_B \to 0$. Also $||a_nb - ab||_A \leq ||a_n - a||_A||b||_A \to 0$, and therefore $ab = c$. This proves that B is a left ideal in $\overline{B}^A$. Furthermore

$$||ab||_B = ||c||_B = \lim_n ||a_nb||_B \leq \lim_n M||a_n||_A||b||_B = M||a||_A||b||_B. \ \square$$

(31.5) <u>Theorem</u>. Let $(B,||\cdot||_B)$ be an abstract Segal algebra in $(A,||\cdot||_A)$. If B is a proper subset of A, then $(B,||\cdot||_B)$ cannot have bounded right approximate units.

<u>Proof</u>. Suppose $(B,||\cdot||_B)$ has right approximate units bounded by $H \geq 1$. Choose $b \in B$. Then there is a sequence $\{u_n\}$ of elements in B with $||u_n|| \leq H$ such that $||bu_n - b||_B \to 0$. Thus

$$||b||_A \leq C||b||_B = C \lim_n ||bu_n||_B \leq C \limsup_n M||b||_A||u_n||_B \leq CMH||b||_A.$$

But this estimate implies that the norms $||\cdot||_A$ and $||\cdot||_B$ are equivalent on B which is a contradiction since B is a proper dense subset of A. □

Now assuming that $(B,||\cdot||_B)$ is an abstract Segal algebra in $(A,||\cdot||_A)$, we relate the closed ideals of B to those of A. The most comprehensive results in this direction are obtained in the presence of right approximate units.

(31.6) Lemma. Let $(B,||\cdot||_B)$ be an abstract Segal algebra in $(A,||\cdot||_A)$. Assume that $(A,||\cdot||_A)$ has right approximate units. If I is a closed right ideal of $(A,||\cdot||_A)$ then $I \cap B$ is a closed right ideal of $(B,||\cdot||_B)$ and $I = \overline{I \cap B}^A$.

Proof. Let I be a closed right ideal of $(A,||\cdot||_A)$. Then $I \cap B \neq B$ since otherwise $I = \overline{I}^A = \overline{B}^A = A$ in contradiction to $I \neq A$. Thus $I \cap B$ is a proper right ideal of B. To see that $I \cap B$ is closed in $(B,||\cdot||_B)$ let $\{b_n\} \subset I \cap B$ and $b \in B$ such that $||b_n - b||_B \to 0$. Then $||b_n - b||_A \leq C||b_n - b||_B \to 0$. Since I is closed in $(A,||\cdot||_A)$ it follows that $b \in I$. Hence $b \in I \cap B$ and so $I \cap B$ is closed in $(B,||\cdot||_B)$.

Obviously, $\overline{I \cap B}^A \subset \overline{I}^A = I$. To establish the reverse inclusion choose $a \in I$ and let $\varepsilon > 0$ be given. Since $(A,||\cdot||_A)$ has right approximate units there exists an element $u \in A$ such that

$$||au - a||_A < \varepsilon.$$

Since B is dense in $(A,||\cdot||_A)$ there is an element $b \in B$ such that

$$||b - u||_A < \varepsilon/||a||_A.$$

Then

$$||ab - a||_A \leq ||ab - au||_A + ||au - a||_A < 2\varepsilon.$$

The element ab belongs to I, since I is a right ideal, and to B, since B is a left ideal in A. Thus $a \in \overline{I \cap B}^A$ and so $\overline{I \cap B}^A = I$. □

(31.7) <u>Lemma</u>. Let $(B, ||\cdot||_B)$ be an abstract Segal algebra in $(A, ||\cdot||_A)$. Assume that $(B, ||\cdot||_B)$ has right approximate units. If J is a closed right ideal of $(B, ||\cdot||_B)$ then $\overline{J}^A$ is a closed right ideal of $(A, ||\cdot||_A)$ and $J = \overline{J}^A \cap B$.

<u>Proof</u>. Let J be a closed right ideal of $(B, ||\cdot||_B)$. Then $\overline{J}^A$ is a closed subspace of $(A, ||\cdot||_A)$. Clearly $J \subset \overline{J}^A \cap B$. To establish the reverse inclusion choose $b \in \overline{J}^A \cap B$ and let $\varepsilon > 0$ be given. Since $(B, ||\cdot||_B)$ has right approximate units there exists an element $u \in B$ such that

$$||bu - b||_B < \varepsilon.$$

Choose $\{b_n\} \subset J$ such that $||b_n - b||_A \to 0$. Then

$$||b_n u - bu||_B \leq M||b_n - b||_A ||u||_B \to 0.$$

Thus we can choose a positive integer m such that

$$||b_m u - bu||_B < \varepsilon.$$

Hence

$$||b_m u - b||_B \leq ||b_m u - bu||_B + ||bu - b||_B < 2\varepsilon$$

with $b_m u \in J$. Since J is closed in $(B,||\cdot||_B)$ it follows that $b \in J$ and so $J = \overline{J}^A \cap B$. Furthermore, $\overline{J}^A$ is a proper subset of A since otherwise $J = \overline{J}^A \cap B = A \cap B = B$ in contradiction to $J \neq B$.

Finally, if $b \in \overline{J}^A$ and $a \in A$ then there are sequences $\{b_n\} \subset J$ and $\{a_n\} \subset B$ such that $||b_n - b||_A \to 0$ and $||a_n - a||_A \to 0$. Then $b_n a_n \in J$ for all n and $||b_n a_n - ba||_A \to 0$. Hence $ba \in \overline{J}^A$ and so $\overline{J}^A$ is a right ideal of A. □

(31.8) <u>Theorem</u>. Let $(B,||\cdot||_B)$ be an abstract Segal algebra in $(A,||\cdot||_A)$. Assume that $(A,||\cdot||_A)$ and $(B,||\cdot||_B)$ have right approximate units. Then the mapping

$$I \to I \cap B$$

is bijective from the set of all closed right (two-sided) ideals in $(A,||\cdot||_A)$ onto the set of all closed right (two-sided) ideals in $(B,||\cdot||_B)$. The inverse mapping is

$$J \to \overline{J}^A.$$

<u>Proof</u>. If I is a closed right ideal in $(A,||\cdot||_A)$ then the composition

$$I \to I \cap B \to \overline{I \cap B}^A = I$$

is the identity map.

If J is a closed right ideal in $(B,||\cdot||_B)$ then the composition

$$J \to \overline{J}^A \to \overline{J}^A \cap B = J$$

is also the identity map. Thus the mapping $I \to I \cap B$ is bijective and

$J \to \overline{J}^A$ is its inverse. The statement for two-sided ideals follows from the corresponding statement for right ideals since the mappings $I \to I \cap B$ and $J \to \overline{J}^A$ carry closed two-sided ideals into closed two-sided ideals. □

(31.9) <u>Theorem</u>. Let $(B, ||\cdot||_B)$ be an abstract Segal algebra in $(A, ||\cdot||_A)$. Assume that $(A, ||\cdot||_A)$ and $(B, ||\cdot||_B)$ have right approximate units. Let I be a closed two-sided ideal in $(A, ||\cdot||_A)$ and $I \cap B$ be the corresponding closed two-sided ideal in $(B, ||\cdot||_B)$. Then $(I, ||\cdot||_A)$ has right approximate units if and only if $(I \cap B, ||\cdot||_B)$ has right approximate units.

<u>Proof</u>. Assume $(I, ||\cdot||_A)$ has right approximate units. Choose $b \in I \cap B$ and let $\varepsilon > 0$. Since $(B, ||\cdot||_B)$ has right approximate units there is an element $v \in B$ such that

$$||bv - b||_B < \varepsilon.$$

By assumption there is an element $a \in I$ such that

$$||ba - b||_A < \varepsilon/||v||_B.$$

Then

$$\begin{aligned} ||bav - b||_B &\le ||bav - bv||_B + ||bv - b||_B \\ &\le M||ba - b||_A ||v||_B + ||bv - b||_B \\ &\le (1 + M)\varepsilon, \end{aligned}$$

where $av \in I \cap B$. Thus $(I \cap B, ||\cdot||_B)$ has right approximate units.

Conversely, assume that $(I \cap B, ||\cdot||_B)$ has right approximate units. Choose $a \in I$ and let $\varepsilon > 0$. Since $(A, ||\cdot||_A)$ has right approximate units and B is dense in $(A, ||\cdot||_A)$, there is an element $v \in B$ such that

$$||av - a||_A < \varepsilon.$$

Since $av \in I \cap B$ there is an element $b \in I \cap B$ such that

$$||avb - av||_B < \varepsilon.$$

Then

$$\begin{aligned} ||avb - a||_A &\leq ||avb - av||_A + ||av - a||_A \\ &\leq C||avb - av||_B + ||av - a||_A \\ &\leq (1 + C)\varepsilon, \end{aligned}$$

where $vb \in I$. Thus $(I, ||\cdot||_A)$ has right approximate units. □

We shall give a reformulation of this proof employing the <u>method of linear functionals</u>. Observe that a normed algebra A has right approximate units iff and only if for each $a \in A$, a is in the closed linear subspace $\overline{aA}$ of A. Given $f \in A^*$, $a \in A$, let $\langle f,a\rangle$ and $\langle a,f\rangle$ be the elements of A^* defined by

$$\langle f,a\rangle(x) = f(ax), \text{ resp., } \langle a,f\rangle(x) = f(xa), \ x \in A.$$

Then an easy application of the Hahn-Banach Theorem shows:

III. MORE ABOUT APPROXIMATE IDENTITIES

A normed algebra A has right approximate units if and only if for each $a \in A$ and $f \in A^*$ with $\langle f,a\rangle(A) = \{0\}$ we have $f(a) = 0$.

Now let J be a closed right ideal in $(B,||\cdot||_B)$ and let $I = \overline{J}^A$ be the corresponding closed right ideal in $(A,||\cdot||_A)$. Assume that $(I,||\cdot||_A)$ has right approximate units. Choose $b \in J$ and let $f \in (J,||\cdot||_B)^*$ such that $\langle f,b\rangle(J) = \{0\}$. Since $(B,||\cdot||_B)$ has right approximate units there is a sequence $\{v_n\}$ of elements in B such that $||bv_n - b||_B \to 0$. Then

$$\langle\langle v_n,f\rangle,b\rangle(J) = \langle v_n,\langle f,b\rangle\rangle(J) = \{0\}$$

since J is a right ideal in B. Since

$$|\langle v_n,f\rangle(x)| = |f(xv_n)| \leq ||f||\cdot||xv_n||_B$$

$$\leq ||f||\cdot M\cdot||x||_A||v_n||_B, \quad \text{for all } x \in J,$$

$\langle v_n,f\rangle$ is continuous with respect to the norm $||\cdot||_A$ and so can be extended to a continuous linear functional on $(I,||\cdot||_A)$. Then

$$\langle\langle v_n,f\rangle,b\rangle(I) = \{0\}.$$

Since $b \in J \subset I$ and I has right approximate units we obtain

$$\langle v_n,f\rangle(b) = f(bv_n) = 0;$$

hence

$$f(b) = 0.$$

Thus $(J,||\cdot||_B)$ has right approximate units.

Conversely, let I be a closed two-sided ideal in $(A,||\cdot||_A)$ and $J = I \cap B$ be the corresponding closed two-sided ideal in $(B,||\cdot||_B)$. Assume that $(J,||\cdot||_B)$ has right approximate units. Choose $a \in I$ and let $f \in (I,||\cdot||_A)^*$ such that $\langle f,a\rangle(I) = \{0\}$. Since $(A,||\cdot||_A)$ has right approximate units and B is dense in $(A,||\cdot||_A)$, there is a sequence $\{v_n\}$ of elements in B such that $||av_n - a||_A \to 0$. We have $av_n \in I \cap B = J$, since B is a left ideal and I is a right ideal in A, and

$$\langle f,av_n\rangle(I) = \langle\langle f,a\rangle,v_n\rangle(I) = \{0\},$$

since I is also a left ideal. So

$$\langle f,av_n\rangle(J) = \{0\} \quad \text{since} \quad J \subset I.$$

Since $av_n \in J$ and J has right approximate units we obtain

$$f(av_n) = 0;$$

hence

$$f(a) = 0.$$

Thus $(I,||\cdot||_A)$ has right approximate units. □

32. Sums of subspaces.

(32.1) Theorem. Let X be a left Banach A-module. Suppose Y and Z are closed subspaces of X, and suppose that there is a bounded subset U of A such that $UX \subset Y$, $UZ \subset Z$, and for each $y \in Y$, $y \in \overline{Uy}$. Then $Y + Z$ is a closed subspace of X.

Proof. Let $x \in \overline{Y + Z}$. Choose $\varepsilon_n > 0$ such that $\sum_{n=1}^{\infty} \varepsilon_n < \infty$. There exist elements $v_n \in Y + Z$ such that $||x - v_n|| \to 0$ as $n \to \infty$. Choose a subsequence, again denoted by $\{v_n\}$, such that $||v_n - v_{n-1}|| < \varepsilon_n$ for $n \geq 2$. Put $x_1 = v_1$, $x_n = v_n - v_{n-1}$ for $n \geq 2$. Then $||x_n|| < \varepsilon_n$ for $n \geq 2$, and $x = \Sigma\, x_n$.

Since $x_n \in Y + Z$, there exist $y_n \in Y$, $z_n \in Z$, such that $x_n = y_n + z_n$. Now choose $u_n \in U$ such that

$$||y_n - u_n y_n|| < \varepsilon_n \quad (n \geq 1).$$

Then $x_n = \tilde{y}_n + \tilde{z}_n$, where $\tilde{y}_n = y_n - u_n y_n + u_n x_n$ is in Y and $\tilde{z}_n = z_n - u_n z_n$ is in Z. Let $H \geq 1$ be a bound for U. Then

$$||\tilde{y}_n|| \leq ||y_n - u_n y_n|| + ||u_n x_n|| \leq (1 + MH)\varepsilon_n \quad (n \geq 2)$$

and so

$$||\tilde{z}_n|| = ||x_n - \tilde{y}_n|| \leq ||x_n|| + ||\tilde{y}_n|| \leq (2 + MH)\varepsilon_n \quad (n \geq 2).$$

Since Y and Z are closed subspaces of a Banach space, they are complete. Hence $\tilde{y} = \sum_{n=1}^{\infty} \tilde{y}_n$ is in Y and $\tilde{z} = \sum_{n=1}^{\infty} \tilde{z}_n$ is in Z. Therefore $x = \tilde{y} + \tilde{z}$ is in $Y + Z$. □

(32.2) Corollary. Suppose that L and R are closed left and right ideals, respectively, in a Banach algebra A. If L has bounded right approximate units or if R has bounded left approximate units, then $R + L$ is a closed linear subspace of A.

Proof. Assume R has left approximate units bounded by $H \geq 1$.

Consider A as a left module over itself. Now apply the theorem with $X = A$, $Y = R$, $Z = L$, and $U = \{u \in R: ||u|| \leq H\}$. □

33. Weak approximate identities and Arens products.

(33.1) Definition. Let A be a Banach algebra. A bounded net $\{e_\lambda\}_{\lambda\in\Lambda}$ of elements in A is called a bounded weak left approximate identity if

$$\lim_{\lambda} f(e_\lambda x) = f(x)$$

for every $x \in A$ and $f \in A^*$.

Bounded weak "right" and weak "two-sided" approximate identities are defined similarly. Clearly, every bounded left approximate identity for A is a bounded weak left approximate identity for A. It is interesting to note that if A has a bounded weak left approximate identity $\{e_\lambda\}_{\lambda\in\Lambda}$, then A also has a bounded left approximate identity. Indeed, let H be the convex hull of $\{e_\lambda\}_{\lambda\in\Lambda}$ and let $\varepsilon > 0$. Since $e_\lambda x \to x$ in the weak topology of A for each x in A, there is an element $u \in H$ such that $||x - ux|| < \varepsilon$. Now apply Theorem (9.3) to obtain a bounded left approximate identity for A. Summarizing we have:

(33.2) Proposition. If a Banach algebra A has a bounded weak left approximate identity, then A has a bounded left approximate identity, and conversely.

The notion of weak approximate identity was originally introduced for the study of the second dual A^{**} of a Banach algebra A.

Given $x \in A$, let $\hat{x}$ denote its canonical image in A^{**}, i.e., $\hat{x}(f) = f(x)$ for all $f \in A^*$. Then A^{**} is a Banach space and $x \to \hat{x}$ is an isometric linear isomorphism of A <u>into</u> A^{**}. Richard Arens defined a product on A^{**} under which A^{**} becomes a Banach algebra and $x \to \hat{x}$ is an isometric isomorphism of A into A^{**}.

For $F \in A^{**}$ and $f \in A^*$, define an element $Ff \in A^*$ by

$$(Ff)(x) = F(fx) \qquad (x \in A).$$

Next, for $F, G \in A^{**}$, define an element $FG \in A^{**}$ by

$$(FG)(f) = F(Gf) \qquad (f \in A^*).$$

It is routine to verify that A^{**} with this multiplication, called the <u>Arens product</u>, is a Banach algebra.

Since the product in A^{**} is obtained by extending the product in A by weak*-continuity one variable at a time, it follows that if 1 is a left (right) identity for A, then the canonical image $\hat{1}$ is a left (right) identity for A^{**}.

(33.3) <u>Theorem</u>. Let A be a Banach algebra. Then A has a bounded right approximate identity if and only if the Banach algebra A^{**} has a right identity.

<u>Proof</u>. If A has a bounded right approximate identity $\{e_\lambda\}_{\lambda\in\Lambda}$, then $\{||\hat{e}_\lambda||\}_{\lambda\in\Lambda}$ is bounded and hence has a weak*-cluster point E (Alaoglu's theorem). Now, for $a \in A$, $f \in A^*$, we have $\hat{e}_\lambda(fa) = (fa)(e_\lambda) = f(ae_\lambda) \to f(a)$, and so $E(fa) = f(a)$. Thus $Ef = f$ $(f \in A^*)$; so for all $f \in A^*$

and $F \in A^{**}$ we have $(FE)(f) = F(Ef) = F(f)$, i.e. $FE = F$ for all $F \in A^{**}$.

Conversely, suppose A^{**} has a right identity E. By Goldstine's theorem [76, Theorem V 4.5], the canonical image in A^{**} of the ball of radius $||E||$ in A is weak*-dense in the ball of radius $||E||$ in A^{**}. Hence there is a net $\{e_\lambda\}_{\lambda\in\Lambda}$ in A with $||e_\lambda|| = ||\hat{e}_\lambda|| \leq ||E||$, $\lambda \in \Lambda$, such that $\hat{e}_\lambda(f) \to E(f)$, $f \in A^*$, in the weak*-topology. So $\hat{e}_\lambda(fa) \to E(fa)$ for $a \in A$, $f \in A^*$. However $\hat{e}_\lambda(fa) = (fa)(e_\lambda) = f(ae_\lambda)$ and $E(fa) = (Ef)(a) = \hat{a}(Ef) = (\hat{a}E)(f) = \hat{a}(f) = f(a)$. Consequently, $f(ae_\lambda) \to f(a)$, $a \in A$, $f \in A^*$, and the theorem follows from (33.2). □

Let A_r be the reversed algebra of the Banach algebra A with the reversed product $a \bullet b = ba$. One can construct the Arens product $F \circ G$ in $(A_r)^{**}$. In general $F \circ G$ is <u>not</u> the reversed product of FG, i.e., in general $F \circ G \neq GF$. The Arens product is said to be <u>regular</u> if $F \circ G = GF$ for all G, F in A^{**}.

(33.4) <u>Corollary</u>. Suppose the Arens product in A^{**} is regular. Then A^{**} has an identity if and only if A has a bounded two-sided approximate identity.

<u>Proof</u>. Apply Theorem (33.3) and Proposition (2.6). □

A Banach algebra A is said to be <u>reflexive</u> if $A = A^{**}$. The results of this section show that if A is reflexive then A admits a right identity if and only if A admits a bounded right approximate identity.

III. MORE ON APPROXIMATE IDENTITIES

34. A theorem on continuous bilinear mappings.

We present several results in this section for algebras with bounded approximate identities. The main result is a general theorem concerning bilinear mappings from a normed A-module into a normed space. The theorem has several immediate corollaries which are interesting. For example if A is a normed space which is also a normed algebra under each of two multiplications $*$ and $\circ$, then these multiplications coincide if and only if there exists an $\alpha > 0$ such that $||a \circ b|| \leq \alpha ||a * b||$ for a, b in A. This is a condition for the Arens product in A^{**} to be regular. By taking $*$ to be multiplication of a Banach algebra and $\circ$ to be the reversed product, we obtain a condition which implies that A is commutative. Other applications are concerned with conditions under which a bilinear mapping between two algebras is a homomorphism, when an element lies in the center of an algebra, and when a one-dimensional subspace of an algebra is a right ideal.

(34.1) Theorem. Let A be a normed algebra with bounded approximate identity $\{e_\lambda\}_{\lambda \in \Lambda}$, X an essential left normed A-module, and Y a normed space. If ψ is a continuous bilinear mapping of $A \times X$ into Y, then $\psi(a,x) = \lim_\lambda \psi(e_\lambda, ax)$ for $a \in A$, $x \in X$, if and only if there is a real constant $\beta > 0$ such that

$$||\psi(a,x)|| \leq \beta ||ax|| \tag{1}$$

for all $a \in A$, $x \in X$.

Proof. Assume that (1) is true for some constant $\beta > 0$. Since ψ

extends by continuity to the completions of the spaces involved, we may assume that all three spaces are complete. If A does not have an identity, let A_1 denote, as usual, the algebra A with the identity adjoined. Define $\psi_1 : A_1 \times X \to Y$ by the equation $\psi_1(1,x) = \lim_\lambda \psi(e_\lambda,x)$ and linearity. The limit exists by (1) and the fact that X is essential. It follows that we may assume that A has an identity 1 with $||1|| = 1$, and prove that $\psi(a,x) = \psi(1,ax)$.

Now, let $f \in Y^*$. Given $a \in A$, $x \in X$ let F be defined on the complex numbers C by

$$F(z) = (f\circ\psi)(\exp(-za), \exp(za)x).$$

Then F is an entire function and, for $z \in C$,

$$\begin{aligned}|F(z)| &\le ||f||\cdot||\psi(\exp(-za), \exp(za)x)|| \\ &\le \beta||f||\cdot||x||.\end{aligned}$$

By Liouville's theorem F is constant, and so the coefficient of z in the power series expansion of F is zero, i.e., $f(\psi(1,ax) - \psi(a,x)) = 0$. Since f was arbitrary, $\psi(1,ax) = \psi(a,x)$. The converse of the theorem is obvious. □

(34.2) <u>Corollary</u>. Let A be a normed space which is a normed algebra with bounded approximate identity $\{e_\lambda\}_{\lambda\in\Lambda}$ for each of two multiplications $*$ and $\circ$. These multiplications coincide if and only if there exists a real constant $\beta > 0$ such that $||a\circ b|| \le \beta||a*b||$ for $a, b \in A$.

<u>Proof</u>. Take $Y = X = A$ where A has multiplication $*$, and set $\psi(a,b) = a\circ b$. □

(34.3) Corollary. (LePage; Hirschfeld-Żelazko). A normed algebra A with bounded approximate identity is commutative if and only if there exists $\beta > 0$ such that $||ba|| \leq \beta||ab||$ for $a, b \in A$. This holds in particular if $||a|| \leq \beta|a|_\sigma$ for $a \in A$, where $|.|_\sigma$ denotes the spectral radius.

Proof. The first statement is immediate from (34.2) on taking $*$ to be the multiplication in A and $\circ$ its opposite, i.e., $a \circ b = b * a$. If the second inequality holds, then for $a, b \in A$, $||ba|| \leq \beta|ba|_\sigma = \beta|ab|_\sigma \leq \beta||ab||$. □

(34.4) Corollary. Let A be a normed algebra with bounded approximate identity and $f \in A^*$. If there exists $\beta > 0$ such that $|f(a)| \leq \beta|a|_\sigma$ for $a \in A$, then $f(ab) = f(ba)$ for $a, b \in A$.

Proof. Set $\psi(a,b) = f(ba)$. Then the argument of (34.3) shows that $|\psi(a,b)| < \beta||ab||$ so that Theorem (34.1) applies. □

(34.5) Corollary. Let A and B be normed algebras with identities 1 and $1'$ respectively. Assume that T is a continuous linear mapping of A into B with $T(1) = 1'$. Then T is a homomorphism if and only if there exists $\beta > 0$ such that $||T(a)T(b)|| \leq \beta||ab||$ for $a, b \in A$.

Proof. In (34.1) take $X = A$ and $Y = B$, and set $\psi(a,b) = T(a)T(b)$. □

(34.6) Corollary. Let A be a normed algebra with a bounded approximate identity $\{e_\lambda\}_{\lambda \in \Lambda}$, and let X_1 and X_2 be essential left normed A-modules. Then a continuous linear mapping $T : X_1 \to X_2$ is A-linear, i.e., $T(ax) = aT(x)$ for $a \in A$, $x \in X_1$, if and only if there is a $\beta > 0$ such that $||aT(x)|| \leq \beta||ax||$ for $a \in A$, $x \in X_1$.

Proof. In (34.1) take $X = X_1$, $Y = X_2$, and put $\psi(a,x) = aT(x)$. Then $aT(x) = \psi(a,x) = \lim_\lambda e_\lambda T(ax) = T(ax)$. □

(34.7) Corollary. Let A be a normed algebra with identity 1. Let f be a continuous linear functional on A with $f(1) \neq 0$. Assume that $a \in A$ satisfies $||f(x)ay|| \leq \beta||xy||$ for some real $\beta > 0$ and all x, y in A. Then the subspace $\{\lambda a: \lambda \in C\}$ is a right ideal of A.

Proof. Take $X = Y = A$, and set $\psi(x,y) = f(x)ay$. Then (34.1) gives $f(1)axy = f(x)ay$. Set $y = 1$; then $f(1)ax = f(x)a$. □

(34.8) Corollary. Let A be a normed algebra with bounded approximate identity $\{e_\lambda\}_{\lambda\in\Lambda}$. Then an element $a \in A$ is in the center $Z(A)$ of A if and only if there exists $\beta > 0$ so that $||xay|| \leq \beta||xy||$ for x, y in A.

Proof. In (34.1), take $X = Y = A$, and put $\psi(x,y) = xay$ for x, y in A. Then (34.1) says that $xay = \lim_\lambda e_\lambda axy = axy$. Therefore, $xa = \lim_\lambda xae_\lambda = \lim_\lambda axe_\lambda = ax$. □

35. A majorization theorem for C^*-algebras.

Let A be a C^*-algebra and B a C^*-subalgebra of A. If a is a positive element in A, does there exist a positive element b in B such that $b \geq a$? We shall see in this section that if the subalgebra B has a suitable approximate identity, this is indeed the case. We shall, in addition, be interested in comparing the norms of the elements a and b.

III. MORE ON APPROXIMATE IDENTITIES

If $\{p_\alpha\}$ is an increasing net of projections in a C^*-algebra, then it is well-known and easy to prove that $||a - p_\alpha a|| + ||a - ap_\alpha|| \to 0$ if and only if $||a - p_\alpha a p_\alpha|| \to 0$ for all $a \in A$.

(35.1) <u>Theorem</u>. Let A be a C^*-algebra and B a C^*-subalgebra of A which contains a positive, increasing approximate identity $\{e_\lambda\}_{\lambda\in\Lambda}$ bounded by 1 for A. Then, given $a_o \geq 0$ in A and $\varepsilon > 0$, there is an element $b \geq 0$ in B with $b \geq a_o$ and $||b|| \leq ||a_o|| + \varepsilon$.

<u>Proof</u>. We first note that if the theorem is true for all $a_o \in A$ with $||a_o|| = 1$, then it is true for all $a_o \in A$. Hence we may assume that $||a_o|| = 1$. Given $\varepsilon > 0$ we may choose $\lambda_1 \in \Lambda$ such that

$$||a_o - e_{\lambda_1} a_o e_{\lambda_1}|| < \frac{\varepsilon}{2}.$$

Since $\{e_\lambda\}_{\lambda\in\Lambda}$ is an increasing positive approximate identity bounded by 1 and $||a_o|| = 1$, then $e_{\lambda_1} \geq e_{\lambda_1} a_o e_{\lambda_1}$. Set $a_1 = a_o - e_{\lambda_1} a_o e_{\lambda_1}$. Now choose $\lambda_2 \in \Lambda$ such that

$$||a_1 - e_{\lambda_2} a_1 e_{\lambda_2}|| < \frac{\varepsilon}{2^2}.$$

Since $||(2/\varepsilon)a_1|| \leq 1$, then as above we have

$$e_{\lambda_2} \geq e_{\lambda_2}^2 \geq e_{\lambda_2}((2/\varepsilon)a_1)e_{\lambda_2}.$$

Hence $(\varepsilon/2)e_{\lambda_2} \geq e_{\lambda_2} a_1 e_{\lambda_2}$. Set $a_2 = a_1 - e_{\lambda_2} a_1 e_{\lambda_2}$, and continue by induction to get sequences $\{\lambda_n\}_{n=1}^\infty$ in Λ and $\{a_n\}_{n=0}^\infty$ in A so that $a_n = a_{n-1} - e_{\lambda_n} a_{n-1} e_{\lambda_n}$ for $n > 0$, where a_o is given, $||a_n|| \leq \varepsilon/2^n$

for $n > 0$, and

$$(\varepsilon/2^{n-1})e_{\lambda_n} \geq e_{\lambda_n} a_{n-1} e_{\lambda_n} .$$

Therefore the series $\sum_{n=1}^{\infty} e_{\lambda_n} a_{n-1} e_{\lambda_n}$ is absolutely convergent to a_o. Further,

$$a_o = \sum_{n=1}^{\infty} e_{\lambda_n} a_{n-1} e_{\lambda_n} \leq e_{\lambda_1} + \sum_{n=2}^{\infty} (\varepsilon/2^{n-1}) e_{\lambda_n} ,$$

the right hand side also converging absolutely to an element $b \in B$. Clearly

$$||b|| \leq ||e_{\lambda_1}|| + \sum_{n=2}^{\infty} ||(\varepsilon/2^{n-1})e_{\lambda_n}|| \leq 1 + \varepsilon,$$

so the theorem is proved. □

By utilizing Gelfand theory (see the Appendix) one can show in the commutative case that $||b|| = ||a_o||$ in (35.1). However, in the absence of commutativity this may fail as the following example shows.

(35.2) Example. Let H be a separable Hilbert space and $\{x_n\}_{n=1}^{\infty}$ an orthonormal basis for H. Let A be the algebra of all compact operators on H and $B = \{a \in A$: each x_i is an eigenvector of $a\}$. This means that B is the algebra of compact diagonal operators for the basis $\{x_n\}_{n=1}^{\infty}$. For each $n = 1,2,\ldots$ define q_n to be the orthogonal projection on the subspace spanned by $\{x_1,\ldots,x_n\}$. Then $\{q_n\}_{n=1}^{\infty}$ is is a positive, increasing approximate identity bounded by 1 for A. Also B is the C*-subalgebra of A generated by $\{q_n\}_{n=1}^{\infty}$. Hence the theorem applies.

Now let p be the orthogonal projection on the subspace spanned by any vector $x = \Sigma_{i=1}^{\infty} \alpha_i x_i \in H$, where we assume $\alpha_i \neq 0$ for all i and $||x|| = 1$. To see that the ε condition of the theorem is necessary, suppose there was some $b \in B$, $||b|| = 1$, $b \geq p$. Then

$$\begin{aligned}(bx|x) &= \sum_{n=1}^{\infty} (bx_n|x_n)|\alpha_n|^2 \\ &\geq (px|x) = (x|x) \\ &= \sum_{n=1}^{\infty} |\alpha_n|^2.\end{aligned}$$

This would mean that $(bx_n|x_n) = 1$ for all n, contradicting the compactness of the operator b. □

36. Approximate diagonals.

In this section we introduce briefly the notions of approximate diagonal and amenability for Banach algebras and groups. Our purpose is merely to inform the reader of the connection of these topics with approximate identities and Banach modules. For more complete accounts of the subject we refer the reader to [29, pp. 231-247], [139] and [140].

Let A be a Banach algebra. The projective tensor product $A \otimes_\gamma A$ is the completion of the algebraic tensor product $A \otimes A$ with respect to the greatest cross-norm $||\cdot||_\gamma$. The space $A \otimes_\gamma A$ becomes a Banach A-module if we define $a(b \otimes c) = ab \otimes c$ and $(b \otimes c)a = b \otimes ca$ for a, b, c in A. Let $\pi : A \otimes_\gamma A \to A$ be the continuous linear mapping defined by $\pi(a \otimes b) = ab$ for $a, b \in A$.

(36.1) Definition. An approximate diagonal for a Banach algebra A is a bounded net $\{m_\alpha\}$ in $A \otimes_\gamma A$ such that $m_\alpha a - am_\alpha \to 0$ and $\pi(m_\alpha)a \to a$ for all $a \in A$, in the respective norm topologies.

Clearly an approximate diagonal, if one exists, can be chosen from $A \otimes A$. If $\{m_\alpha\}$ is an approximate diagonal for A, then $\{\pi(m_\alpha)\}$ is a bounded two-sided approximate identity in A. If A has an identity 1, then $\pi(m_\alpha)a \to a$ for all $a \in A$ if and only if $\pi(m_\alpha) \to 1$.

(36.2) Example. Let A be the Banach algebra c_o. For $i = 0,1,\ldots$ let e_i be the element of c_o with $(e_i)_j = \delta_{ij}$. Then $m_n = e_o \otimes e_o + \ldots + e_n \otimes e_n$ is an approximate diagonal. Indeed, since

$$\frac{1}{n} \sum_p [\sum_j (\exp 2\pi ijp/n)e_j] \otimes [\sum_k (\exp 2\pi ikp/n)^{-1} e_k]$$

$$= \frac{1}{n} \sum_{j,k} (\sum_p \exp 2\pi i(j-k)p/n)e_j \otimes e_k = m_{n-1}$$

where all sums range over $0,1,\ldots,n-1$ and $||\sum_j (\exp 2\pi ijp/n)^{\pm 1} e_j|| = 1$, we see that $||m_{n-1}|| \leq 1$. Clearly $\pi(m_n) = e_o + e_1 + \ldots + e_n$ and so $\pi(m_n)a \to a$ as $n \to \infty$ for all $a \in A$. Also $m_n a = am_n$, so that $\{m_n\}$ is an approximate diagonal. □

(36.3) Definition. A Banach algebra with identity is said to be amenable if it has an approximate diagonal. A group G is called amenable if the group algebra $\ell^1(G)$ is an amenable Banach algebra.

Let A be a Banach algebra with identity and X a Banach A-bimodule. Form the dual Banach A-bimodule X^*, where the module products are given by $(a\cdot f)(x) = f(xa)$, $(f\cdot a)(x) = f(ax)$ for $a \varepsilon A$, $f \varepsilon X^*$, $x \varepsilon X$.
A _bounded_ X-_derivation_ is a bounded linear mapping D of A into X such that

$$D(ab) = (Da)b + a(Db)$$

for all $a, b \varepsilon X$. The set of bounded X-derivations will be denoted by $Z^1(A,X)$; it is a linear subspace of the space of all bounded linear mappings from A into X. For $x \varepsilon X$, let $\delta_x : A \to X$ be defined by

$$\delta_x(a) = ax - xa$$

for $a \varepsilon A$. Then $\delta_x \varepsilon Z^1(A,X)$; δ_x is called an _inner_ X-_derivation_, the set of which will be denoted by $B^1(A,X)$. $B^1(A,X)$ is a linear subspace of $Z^1(A,X)$ and we will denote by $H^1(A,X)$ the quotient space of $Z^1(A,X)$ modulo $B^1(A,X)$,

$$H^1(A,X) = Z^1(A,X)/B^1(A,X).$$

$H^1(A,X)$ is called the _first cohomology group of_ A with coefficients in X.

One can show [29, p. 243] that a Banach algebra A with identity is amenable if and only if $H^1(A,X) = \{0\}$ for every Banach A-bimodule X. Examples of amenable Banach algebras are $C(E)$, the algebra of continuous complex-valued functions on a compact Hausdorff space E, and the algebra of of compact operators (with the identity adjoined) on a separable Hilbert space. An example of a non-amenable Banach algebra is provided by the disk algebra, i.e., the algebra of all continuous complex-valued functions defined on $\Delta = \{\lambda \in C: |\lambda| \leq 1\}$ which are holomorphic on the interior of Δ.

An <u>invariant mean</u> on a (discrete) group G is a positive functional M on $\ell^\infty(G)$ such that $M(1) = 1$ and $M(T_g f) = M(f)$ for all $f \in \ell^\infty(G)$, $g \in G$, where T_g is the left translation operator on $\ell^\infty(G)$ defined by $(T_g f)(h) = f(g^{-1}h)$ for all $f \in \ell^\infty(G)$, $g \in G$. It can be shown [29, p. 239] that a group G is amenable if and only if there exists an invariant mean on G (this is the original definition for amenability of groups.) Among the class of amenable groups are the finite groups and the abelian groups [29, p. 241].

As an example, we note here that the group Z of integers is amenable (see [142]). Indeed, let F be a free ultrafilter on Z^+ and define $M(a) = \lim_{n \in F} (1/(2n+1)) \sum_{i=-n}^{n} a_i$. Then $M(1) = 1$ and since

$$\left|\frac{1}{2n+1} \sum_{i=-n}^{n} a_i - \frac{1}{2n+1} \sum_{i=-n}^{n} a_{i+m}\right| = \left|\frac{1}{2n+1} \left(\sum_{i=-n}^{-n+m-1} a_i - \sum_{i=n+1}^{n+m} a_i\right)\right|$$

$$\leq \frac{1}{2n+1} ||a||_\infty \to 0$$

as $n \to \infty$ we have

$$\lim_{n \in F} \frac{1}{2n+1} \sum_{i=-n}^{n} a_i = \lim_{n \in F} \frac{1}{2n+1} \sum_{i=-n}^{n} a_{i+m},$$

so M is an invariant mean on Z. An interesting alternative description of

M is to take a state (positive functional of norm one) ρ on the C^*-algebra $\ell^\infty(Z^+)/c_o$ and let $M(a)$ be the value of ρ on the coset in $\ell^\infty(Z^+)/c_o$ of the sequence

$$b_n = \frac{1}{2n+1} \sum_{i=-n}^{n} a_i .$$

Since the sequence $\{b'_n\}$ associated with a translate of $\{a_n\}$ satisfies $b - b' \in c_o$, we see that $\rho(b) = \rho(b')$.

An example of a non-amenable group is provided by the free group on two or more generators.

If G is a locally compact topological group, amenability is defined in terms of the Banach algebra $L^\infty(G)$ instead of $\ell^\infty(G)$, see Greenleaf [113].

NOTES AND REMARKS

Chapter I.1. The concept of a bounded approximate identity goes back to the earliest studies of the group algebra $L^1(G)$ [282]. Approximate identities in $L^1(G)$ are discussed in a systematic way in A. Weil's book [281], especially pp. 52, 79-80, and 85-86, and they have since become a standard tool in harmonic analysis [131], §28 (called _bounded approximate unit_); [80], Chapter III (called _approximate identity_); [147], Chapter I (called _summability kernel_); [220], §6 (called _bounded multiple units_).

For applications to group representations it soon became necessary to consider C^*-algebras without identity element. In order to discuss such algebras I. E. Segal constructed for every C^*-algebra an approximate identity bounded by one [243], Lemma 1.1. Many results about a C^*-algebra without an identity element can be obtained by embedding such an algebra in a C^*-algebra with identity element, called _adjunction of an identity_ (see [222], Lemma (4.1.13) or [70], 1.3.8). But some problems, especially those which involve approximate identities, are not susceptible to this approach. Therefore, the approximate identity is the main tool in Dixmier's book [70] (called _unité approchée_ in the earlier French editions) to carry through all the basic theory of C^*-algebras.

In recent years many results known for C^*-algebras or for Banach algebras with identity have been extended to algebras with an approximate identity. The mainspring for much of this work was the Cohen-Hewitt factorization theorem for Banach modules [131], §32.

The elementary results in I.1 are very familiar. For Proposition (1.2) see H. Reiter [220], §6; for Lemma (1.4) see P. G. Dixon [71], Lemma (2.1); for Proposition (1.5) see J. Dixmier [70], 1.7.2.

2. The results relating left, right, and two-sided approximate identities are due to P. G. Dixon [71]. Our Lemma (2.1) is a reformulation of Dixon's Lemma (3.1) which contained an ambiguity in its statement. The authors thank Professor C. R. Combrink for his insights on this lemma. For Proposition (2.5) see [71], Proposition (4.4) and also C. R. Warner and R. Whitley [280], p. 279.

3. It is well known that if a normed algebra A has an identity then there is an equivalent algebra norm on A such that the identity has norm one [30], I.2.1, pp. 13-14; [152], p. 52. The analogue for approximate identities has been studied by P. G. Dixon [71], [72]. For Lemma (3.2) see F. F. Bonsall and J. Duncan [27], p. 21; for Theorem (3.3) see also W. B. Johnson [144], p. 309. Theorem (3.4) can be found in Dixon [72] as Theorem (2.3), and Theorem (3.5) is due to A. M. Sinclair [250], Theorem 8 (the version given here is a <u>nonseparable</u> generalization of Sinclair's theorem to <u>normed</u> algebras, see [72], Theorem (2.4)).

4, 5, 6. The results of these three sections have been taken from Dixon [72].

7. The result about the existence of approximate identities in quotient algebras can be found in H. Reiter [220], §7, Lemma 2 and Lemma 3.

Problem. Let A be a normed algebra and let I be a closed two-sided ideal in A. If I and A/I have left approximate identities, does A have a left approximate identity?

Example (7.2) is due to P. G. Dixon (private communication).

8. For the general theory of tensor products of normed linear spaces see R. Schatten's book [241] or A. Grothendieck's monograph [119]; for the tensor product of Banach algebras and Banach modules see [29], [102], [103], [104], [172], [120].

The first result concerning identities in $A \otimes_{\alpha} B$ was that of B. R. Gelbaum [100], Theorem 4: Let A and B be commutative semisimple Banach algebras. Then $A \otimes_{\gamma} B$ has an identity if and only if A and B have identities. L. J. Lardy and J. A. Lindberg proved this result for any "spectral tensor norm" [163]; for an elementary proof see R. J. Loy [185].

The fact that $A \otimes_{\alpha} B$ for any admissible norm $||\cdot||_{\alpha}$ has a bounded approximate identity if A and B have bounded approximate identities seems to be well known; it is used implicitly by K. B. Laursen in [173], Theorem 2.2. The first systematic study of approximate identities in tensor products of Banach algebras was made by R. J. Loy [185]. D. A. Robbins [230] and J. R. Holub [133] improved his results. The short proof of Theorem (8.2) in the text was communicated to us privately by B. E. Johnson.

9. The need for the more general concept of approximate units in a normed algebra first arose in connection with the study of Wiener-Ditkin sets [218], Chapter 7, §4, especially §4.10, and became more apparent in

H. Reiter's lecture notes [220].

Proposition (9.2) was motivated by M. Altman's study of contractors [8], [10], [11]. For a systematic study of contractors see Altman's recent book [13] on the subject.

The most important and very useful result is the equivalence of the existence of bounded approximate units and the existence of a bounded approximate identity in any normed algebra (Theorem (9.3) and Theorem (9.4)). The present sharp result was obtained successively by H. Reiter [219], [220], §7, Lemma 1; M. Altman [10], Lemma 1; and J. Wichmann [284].

The study of pointwise-bounded approximate units was taken up by Wichmann in [147]; for Theorem (9.7), see also T. S. Liu, A. van Rooij, and J. K. Wang [182], Lemma 12.

Problem. Let A be a Banach algebra with pointwise-bounded left approximate units. Does A have a bounded left approximate identity?

Problem. Let A be a Banach algebra with left approximate units. Does A have a left approximate identity?

Problem. Does Theorem (9.3) generalize to Banach A-modules?

(Example: Let X be a Banach space and let B'(X) be the Banach algebra of compact operators on X. Then for every $x \in X$ and every $\varepsilon > 0$ there exists a compact operator $u \in B'(X)$, depending on x and ε, with $||u|| \leq 1$ such that $||x - ux|| < \varepsilon$. Does there exist a bounded net $\{e_\lambda\}_{\lambda\in\Lambda}$ of compact operators on X such that $\lim_{\lambda\in\Lambda} e_\lambda x = x$ for all $x \in X$?).

The appropriate reformulation of statements and proofs concerning approximate identities for approximate units is left to the reader.

For approximate units in Segal algebras and the method of linear functionals see Chapter III, §32.

10. First results about approximate identities in normed algebras which do not consist entirely of topological zero divisors appeared in R. J. Loy [185], Proposition 1 and Proposition 3. These results have been extended in this section by the second author, with the aim to solve for such algebras the problems mentioned above about approximate units. Some illustrative examples and counterexamples would be desirable.

11. This section was motivated by the interesting paper [199] of J. K. Miziolek, T. Müldner, and A. Rek on topologically nilpotent algebras. Proposition (11.6) extends [199], Proposition 2.4.

12. The study of C^*-algebras without an identity element is more than just a mildly interesting extension of the case of a C^*-algebra with identity. The main tool is the approximate identity which such algebras have. In fact, as the proof of Theorem (12.4) shows, in any left ideal I of a C^*-algebra there is an increasing net $\{e_\lambda\}_{\lambda\in\Lambda}$ of positive operators bounded by one such that $\{e_\lambda\}_{\lambda\in\Lambda}$ is a right approximate identity in the norm closure $\overline{I}$ of I. The construction of such an approximate identity is due to I. E. Segal [243], Lemma 1.1. It was proved by J. Dixmier, Traces sur les C^*-algèbres, Ann. Inst. Fourier 13 (1963), 219-262, that this approximate identity is increasing; see also [70], 1.7.

To see the importance of approximate identities in the extension of the Gelfand-Naimark theorem for noncommutative C^*-algebras (with weak norm condition $||x^*x|| = ||x^*||\cdot||x||$) with identity to the case of C^*-algebras without identity see R. S. Doran and J. Wichmann [73]. Also, see the original paper by B. J. Vowden, On the Gelfand-Naimark theorem, J. London Math. Soc. 42 (1967), 725-731.

In the case of a commutative C^*-algebra $C_o(S)$ the following question arises: what do restrictions on the approximate identity imply about the spectrum S of $C_o(S)$ and vice versa? Along this line, Proposition (12.1) characterizes compactness of the spectrum; the characterization of σ-compactness of S in Proposition (12.2) is due to H. S. Collins and J. R. Dorroh [52], Theorem 4.1. We will discuss later (§28) the characterization of paracompactness (every open cover has a locally finite open refinement) of S in terms of approximate identities. It is interesting to ask similar questions for an arbitrary C^*-algebra A and the space $S = \text{Prim}(A)$ of primitive ideals furnished with the Jacobson topology (or the space $\hat{A}$ of classes of nonzero irreducible representations of A) [70], Chapter 3.

Recall that a *-algebra is symmetric if for every x, the spectrum $\sigma(x^*x)$ is a subset of the nonnegative reals.

Problem. Does there exist a commutative, semisimple, symmetric Banach *-algebra with σ-compact carrier space but no bounded sequential approximate identity?

The construction of a sequential increasing abelian approximate identity for C^*-algebras with a strictly positive element (see (12.9)) is due to J. F. Aarnes and R. V. Kadison [1]. For an interesting application see G. A. Elliott's paper [83] on a weakening of the axioms for a C^*-algebra.

NOTES AND REMARKS

Remark (12.13) is taken from H. S. Collins and R. A. Fontenot [53], Lemma 4.4. Example (12.14) was constructed by C. A. Akemann [4], Example 2.1. His proof contained several gaps; the corrected version given here is due to B. E. Johnson (private communication).

Recall that the set B of positive functionals in the unit ball of the dual space of a C^*-algebra A is compact and convex in the weak*-topology [70,2.5.5]. Thus the Krein-Milman theorem implies that B is the weak*-closure of the convex linear combinations of its extreme points. These extreme points are called <u>pure states</u>. In [2], p. 531, Akemann gave the following characterization of approximate identities in a C^*-algebra.

<u>Theorem</u> 1. Let A be a C^*-algebra and $\{e_\lambda\}_{\lambda\varepsilon\Lambda}$ an increasing net of positive elements in A such that $f(e_\lambda) \to 1$ for all pure states f in A^*. Then $\{e_\lambda\}_{\lambda\varepsilon\Lambda}$ is an approximate identity for A.

A pure state f of a C^*-algebra A is said to be <u>pure on a closed left ideal</u> I of A if f is pure on the C^*-algebra $I \cap I^*$. The following result due to Akemann [4], Theorem 1.3, extends Theorem 1 above to <u>right</u> approximate identities for <u>left</u> ideals of C^*-algebras.

<u>Theorem</u> 2. Let I be a closed left ideal of a C^*-algebra A. An increasing net of positive elements $\{e_\lambda\}_{\lambda\varepsilon\Lambda}$ in I is a right approximate identity for I if $f(e_\lambda) \to 1$ for every pure state f of A which is pure on I.

Let I be a closed two-sided ideal of a C^*-algebra A. Then, as is well known [70, §12], [239, p. 43] the second dual A^{**} is a von Neumann algebra in a natural manner containing A as a C^*-subalgebra. One identifies A with its image in A^{**}, and I with I^{**} in A^{**} under the double transpose of the inclusion map of I into A. Now $I^{**} = qA^{**}$, where q is a projection in the center of A^{**} [239, proof of 1.17.3, p. 43]. The next characterization of approximate identities, which includes Theorem 1, is due to Akemann and Pedersen [5], p. 123.

Theorem 3. Let A be a C^*-algebra and I a closed two-sided ideal of A. The following are equivalent for a positive increasing net $\{e_\lambda\}_{\lambda\varepsilon\Lambda}$ in I:

(a) $\{e_\lambda\}_{\lambda\varepsilon\Lambda}$ is an approximate identity for I.

(b) $e_\lambda \to q$ in the $\sigma(A^{**},A^*)$-topology.

(c) $f(e_\lambda) \to 1$ for every pure state f of I.

In [2], p. 527, Akemann also introduced the notion of a series approximate identity. A set $\{e_\lambda\}_{\lambda\varepsilon\Lambda}$ of orthogonal projections in a C^*-algebra A such that

$$||a - (\sum_{\lambda\varepsilon K} e_\lambda)a(\sum_{\lambda\varepsilon K} e_\lambda)|| \to 0$$

for all $a \varepsilon A$, as K runs over the finite subsets of Λ directed by inclusion, is called a series approximate identity for A. Utilizing Theorem 1 above we obtain the following alternative definition of a series approximate identity: A set of orthogonal projections $\{e_\lambda\}_{\lambda\varepsilon\Lambda}$ in a C^*-algebra A is a series approximate identity if and only if $\sum_{\lambda\varepsilon\Lambda} f(e_\lambda) = 1$ for each pure state f of A.

13. Approximate identities in the group algebra $L^1(G)$ of a locally compact group G have been studied for many years. Since the two volumes of Hewitt and Ross [130], [131] (especially §28, pp. 87-115) contain an excellent treatment of approximate identities in $L^1(G)$, we have given only a few results here. Theorem (13.1) is due to I. E. Segal; a more general statement appears in [130], Theorem (20.25). Theorem (13.4) appears in Loomis [184], Chapter VI, Theorem 31E, and also in [131], Theorem (20.27). A. Hulanicki and T. Pytlik [134] proved the remarkable result that $L^1(G)$, for any locally compact group G, has an abelian two-sided approximate identity bounded by one (see (12.9) and (19.5) of the text for related results).

Theorem (13.6) concerning central approximate identities in $L^1(G)$ when G is compact was noted by F. P. Greenleaf [112], p. 271. The remark following (13.6) characterizing [SIN]-groups in terms of the existence of a central approximate identity in $L^1(G)$ is due to R. Mosak [201].

Several properties of a locally compact group, including the [SIN]-groups just mentioned, can be characterized by the existence of certain approximate identities in three ideals of $L^1(G)$. For the convenience of the reader we give a summary of these interesting results here.

Let G be a locally compact group, M(G) the Banach algebra of all bounded complex measures, view $L^1(G)$ as the ideal in M(G) consisting of all absolutely-continuous measures with respect to left Haar measure, and set

$$M^o(G) = \{m \in M(G): \int_G dm = 0\}, \quad L^o(G) = M^o(G) \cap L^1(G).$$

NOTES AND REMARKS

Let A be a Banach algebra. We consider the following properties for A:

1°) A has an identity;

2°) A has a bounded left approximate identity $\{e_\lambda\}_{\lambda\varepsilon\Lambda}$;

3°) A has a bounded l.a.i. $\{e_\lambda\}_{\lambda\varepsilon\Lambda}$ in the center Z(A) of A;

4°) A has a bounded l.a.i. $\{e_\lambda\}_{\lambda\varepsilon\Lambda}$ such that $(e_\lambda * f - f) \varepsilon Z(A)$ for every $f \varepsilon A$.

There are <u>six classes</u> of locally compact groups G we are interested in characterizing:

[SIN]: as before, G has a basis of neighborhoods of the identity invariant under all inner automorphisms;

$[FC]^-$: every conjugacy class in G is precompact;

$[FIA]^-$: the group of inner automorphisms is precompact in the group of continuous automorphisms;

$[FD]^-$: the group generated by all commutators $(xyx^{-1}y^{-1})$ is precompact;

[FD]: the group generated by all commutators $(xyx^{-1}y^{-1})$ is finite;

G is <u>amenable</u>: there exists a left invariant mean on $L^\infty(G)$, the space of all measurable essentially bounded functions on G [113], [218].

We know that $L^1(G)$ <u>always</u> satisfies 2°); and 1°) iff G is discrete. As noted above, Mosak [201] proved, $L^1(G)$ satisfies 3°) iff $G \varepsilon$ [SIN]. Reiter [219] proved that $L^o(G)$ satisfies 2°) iff G is amenable; and Kotzmann, Rindler [151] proved that $L^o(G)$ satisfies 3°) iff $G \varepsilon [FIA]^-$.

Using methods of the preceding papers and well known structure theorems H. Rindler has very recently completed the above results to the following list of characterizations:

Theorem. Let G be a locally compact group. Then:

1.	G is discrete	iff	$L^1(G)$ satisfies 1°).
2.	G is compact	iff	$M^o(G)$ satisfies 1°).
3.	G is finite	iff	$L^o(G)$ satisfies 1°).
4.	G is arbitrary locally compact	iff	$L^1(G)$ satisfies 2°).
5.	G is amenable	iff	$M^o(G)$ satisfies 2°).
6.	G is amenable	iff	$L^o(G)$ satisfies 2°).
7.	$G \in$ [SIN]	iff	$L^1(G)$ satisfies 3°).
8.	$G \in [FC]^-$	iff	$M^o(G)$ satisfies 3°).
9.	$G \in [FIA]^-$	iff	$L^o(G)$ satisfies 3°).
10.	G is abelian or discrete	iff	$L^1(G)$ satisfies 4°).
11.	$G \in [FD]^-$	iff	$M^o(G)$ satisfies 4°).
12.	G is abelian or discrete or belongs to [FD]	iff	$L^o(G)$ satisfies 4°).

If the boundedness assumption of the net $\{e_\lambda\}_{\lambda\in\Lambda}$ is dropped then the following remarks apply:

If we only assume that A has left approximate units, then parts 5 and 6 remain open. A counter-example to either 5 or 6 for discrete groups would lead to a nonamenable group not containing the free group with two generators.

For a proof of the above theorem see H. Rindler's paper: Approximate units in ideals of group algebras, Proc. Amer. Math. Soc. 71 (1978), 62-64. It is a gem of organization!

We simply mention here that K. McKennon [194] has introduced the notion of ultra-approximate identity for the specific purpose of studying quasi-multipliers of Banach algebras. It is shown in [194] that the group algebra $L^1(G)$, for G locally compact, always possess an ultra-approximate identity bounded by one. One of the main tools to develop the theory is the Cohen-Hewitt factorization theorem (see below).

The results (13.7), (13.8) and (13.9) about the existence of an identity element in the group C^*-algebra $C^*(G)$ are due to P. Milnes [198].

Chapter II.14. The primary tool in the study of Banach algebras with a bounded approximate identity is the Cohen-Hewitt factorization theorem and its many refinements. For applications to harmonic analysis and to function algebras it is convenient to state and prove these results in the general setting of Banach modules [129], [131].

15. Modules over Banach algebras with approximate identities have been considered by several authors [21], [103], [122], [123], [223]. The functor $X \to X_e$ which maps a Banach A-module X to its essential part X_e has been studied in M. A. Rieffel's important paper on induced Banach representations of Banach algebras and locally compact groups [223].

16. Let $L^1(G)$ be the group algebra of a locally compact group. In 1957 Walter Rudin [234] (see also [235]) considered the question of whether each function in $L^1(G)$ is the convolution of two other functions from the same algebra. He showed that this is the case if G is the additive group of Euclidean n-space or the n-dimensional torus. His methods do not extend

to the case of arbitrary groups, because they utilize the Fourier transform and particular functions in Euclidean n-space. Actually, already in 1939 (as mentioned in the text) R. Salem [240] had published the factorization theorems: $L^1(T)*L^1(T) = L^1(T)$, and $L^1(T)*C(T) = C(T)$ for the circle group T.

Subsequently Paul Cohen [51] observed in 1959 that the essential ingredient in Rudin's argument was the presence of a bounded approximate identity in the algebra $L^1(G)$. He went on to prove that the factorization theorem still held in any Banach algebra with a bounded left approximate identity.

It was quickly recognized that Cohen's result could be extended to a more general setting. This extension to Banach modules over Banach algebras with a bounded left approximate identity was published by E. Hewitt [129] (see also [131], Theorem (32.22)), P. C. Curtis and A. Figá-Talamanca [59], S. L. Gulick, T. S. Liu, and A. C. M. van Rooij [122]; their proofs required no ideas different from those in Cohen's proof and it is the first proof given for Theorem (16.1). This first proof has the advantage that the first factor a of the factorization $z = ay$ has the useful infinite sum representation

$$a = \sum_{k=1}^{\infty} c(1-c)^{k-1} e_k.$$

A new method was found by P. Koosis [150]. In his proof the factor a of the factorization $z = ay$ has the infinite product representation

$$a = \prod_{k=1}^{\infty} (1 - c + ce_k) \quad \text{or} \quad a = \prod_{k=1}^{\infty} (1 + c - ce_k)^{-1}.$$

NOTES AND REMARKS

The second proof given for Theorem (16.1) is M. Altman's version [8], [9] of Koosis' proof. Apparently it had not been noticed that a short argument, as given at the end of the second proof, yields an exact bound for the factor a in the factorization $z = ay$. Another proof of (16.1) was given by I. G. Craw [57]. It is not included in §16, but is another version of Koosis' proof. It does not seem to give an exact bound for the factor a.

The third proof, due to V. Pták [213], is a reformulation of Koosis' proof utilizing a generalized closed graph theorem. As pointed out in MR 47 #831, there appear to be gaps in the proof of Pták's earlier paper [212]. Another proof by Pták [214], Theorem (5.3), depends on a very general induction theorem about systems of sets which generalizes the closed graph theorem.

It seems that the factorization of <u>left</u> Banach modules over a Banach algebra with bounded <u>right</u> approximate identity has not been studied; our results are incomplete (see also Discussion (16.4) in the text).

Several types of factorization have been considered in the literature for a Banach algebra A. We mention a few of them here.

(1) A is <u>Cohen factorable</u> if there exists a constant $d > 0$, such that, for $x \in A$ and $\varepsilon > 0$ there are $y, z \in A$ so that $x = yz$, $||y|| \leq d$, and $||x - z|| < \varepsilon$.

(2) A is <u>factorable</u> if, given $x \in A$ there are $y, z \in A$ so that $x = yz$.

(3) A is <u>weakly factorable</u> if, given $x \in A$ there are $y_1, \ldots, y_n$ in A and $z_1, \ldots, z_n$ in A so that $x = \Sigma_{i=1}^{n} y_i z_i$.

NOTES AND REMARKS

(4) A is feebly factorable if, given $x \in A$, there are y_i, z_i in A $(i = 1,2,\ldots)$ with $\Sigma_{i=1}^{\infty} ||y_i||\cdot||z_i|| < \infty$ so that $x = \Sigma_{i=1}^{\infty} y_i z_i$.

It is clear that (1) => (2) => (3) => (4). In this monograph we have only been concerned with Cohen factorization. For results on some of the other types of factorization see H. C. Wang [278].

For several factorization results in C^*-algebras see D. C. Taylor, [260], [262].

17. The factorization of a null sequence by a common factor was noted by N. Th. Varopoulos [269] in order to prove the continuity of positive functionals on Banach *-algebras with bounded approximate identity (Theorem (17.7)). This fact was noted again by B. E. Johnson [137] who used it to show that every centralizer on a Banach algebra with bounded left approximate identity is a bounded linear map (Theorem (17.8)). For further "automatic continuity" results see the monograph [249] by A. M. Sinclair. Theorem (17.9) and Example (17.11) are due to M. D. Green [110].

The simple device of an auxiliary left Banach A-module was introduced by M. A. Rieffel [223], and also by H. S. Collins and W. H. Summers [54].

The factorization of a compact set by a common factor was noted by J. L. Ovaert [207] and also by I. G. Craw [57].

For further factorization results of large sets see D. C. Taylor [258], and F. D. Sentilles and D. C. Taylor [245].

18. The results of this section are due to G. R. Allan and A. M. Sinclair [7]. A somewhat simpler proof of Theorem (18.1) is also given in [7], p. 33 when the algebra A has an abelian bounded left approxi-

mate identity (which covers most of the algebras the man on the street would use). We have not included it because of space considerations.

V. Pták [216] has proved a theorem which is slightly more general than (18.1); his proof depends on his method of nondiscrete mathematical induction [214].

As an application of (18.1), Allan and Sinclair show that if A is a Banach algebra with bounded left approximate identity and B is a suitable radical Banach algebra then the zero homomorphism is the only continuous homomorphism from A into B. In addition, they observe that the growth condition $||y_j|| \leq \alpha_j^j ||z||$ in (18.1)(v) is essentially the best possible. The open question left in [7] has been shown to be false by P. G. Dixon in his review.

19. The results of this section are due to A. M. Sinclair [250]. Note that the general hypothesis in §19 is that all algebras have bounded <u>two-sided</u> approximate identities bounded by one. The properties of the approximate identity in (19.4) and (19.5) are similar to those of the Poisson kernel listed in E. M. Stein's book [254], p.62. As an important application of his results Sinclair proves the following theorem.

<u>Theorem</u>. Let A be a separable Banach algebra. Then A has a bounded approximate identity bounded by one if and only if there is a homomorphism θ from $L^1(R^+)$ into A such that $\theta(L^1(R^+))\cdot A = A = A\cdot\theta(L^1(R^+))$ and $||\theta|| = 1$.

Recently G. Dales [61] and J. Esterle [86] have independently constructed a discontinuous homomorphism on $C_o(X)$, solving a long standing open question (see also the joint paper [62] by Dales and Esterle). The paper of H. G. Dales, Automatic continuity: A survey, Bull. London Math. Soc., 10 (1978), 129-183, is relevent here too. We take this opportunity to mention that Esterle [86] has proved some interesting factorization results based on "Cohen elements" which we have not included in this monograph.

20. The elaboration of the fundamental factorization result $z = ay$ to $z = f(a)y$, where f is any holomorphic function defined on a sufficiently large disc about $\xi = 1$ and satisfying $f(1) = 1$, is due to P. C. Curtis and H. Stetkaer [60].

21. The results in this section are once again due to A. M. Sinclair [251]. As in §19, it is necessary for the Banach algebras in this section to have bounded two-sided approximate identities bounded by one. The examples constructed in Sections 4, 5, and 6 of the text show that the main theorems of this section (and in §19) fail for algebras with bounded one-sided approximate identities. This raises the question of how close a result to Theorem (21.1) is possible for algebras with a one-sided approximate identity. In [251], Theorem 12, Sinclair notes the following result which is a corollary of Theorem (18.1), and he asks if it is essentially the best possible result.

Theorem. Let A be a Banach algebra with a bounded left approximate left identity bounded by one, let X be a left Banach A-module, let $x \in X$ and let $\mathcal{O}$ be the algebra of (germs of) functions analytic in a neighborhood of the disc $\{\lambda \in \mathbb{C}: |\lambda| \leq 1\}$. Then there is an $a \in A$ with $||a|| \leq 1$

such that $x \in f(a)\cdot X$ for all $f \in \mathcal{O}$ such that $f^{-1}(\{0\}) \cap \{\lambda \in C: |\lambda| \leq 1\} = \{0\}$.

The results of §21 have applications to quasicentral approximate identities (see §29), Banach algebras with countable approximate identities, and to proving the existence of symmetric diffusion semigroups on $L^p(G)$, where G is a metrizable locally compact group [251].

22. The examples of noncommutative factorable Banach algebras without approximate identity are due to W. L. Paschke [208] (and private communication). The semisimple commutative factorable Banach algebra without approximate identity constructed in (22.6) is due to M. Leinert [178], [179].

23. The impossibility of factoring by convolution in various subalgebras of group and measure algebras has been studied by many authors. The material presented on the subalgebra $L^1(G) \cap L^p(G)$ can be found in L. Yap [288]. For further results and additional references see [36], [182], [274], [278], [279], [288].

24. It is natural to ask if the previous factorization results for Banach modules can be generalized to a more general class of topological modules. It has been shown that these results with suitable restrictions on the approximate identity can be extended to Fréchet modules over a Fréchet algebra.

For the general theory of locally m-convex topological algebras see E. Michael's monograph [197].

25. Compare the definitions (25.2) and (26.1) of a bounded approximate identity and a uniformly bounded approximate identity.

Problem. Let A be a Fréchet algebra with a bounded approximate identity. Does A have an approximate identity which is uniformly bounded with respect to a suitable sequence of submultiplicative seminorms ?

26. The extension of the factorization theorem for Banach modules to Fréchet modules over a Fréchet algebra with a uniformly bounded approximate identity was annouced by J. L. Ovaert [207]. As in Craw's paper [57], the proof is only a reformulation of the technique used by Koosis [150] for Banach modules to the context of Fréchet modules. W. H. Summers [256] pointed out that for Fréchet modules over a Banach algebra it is possible to give a relatively simple proof based on an application of the Cohen-Hewitt factorization theorem for Banach modules to a suitable auxiliary Banach module. For further factorization results see M. K. Summers [255]. Example (26.5) of a non-normable Fréchet algebra with a uniformly bounded approximate identity is well known [207], [57]; for a discussion of Beurling algebras see [218], [219]. Example 4.2 in [255] is another example of such an algebra.

Example (26.5) of an essential non-metrizable complete locally convex module over a Banach algebra with a bounded approximate identity which cannot be factored is due to W. H. Summers [256]. For further information about the strict topology β, also called the Buck topology, see [23], [33], [53], [245], [246], [260].

NOTES AND REMARKS

Chapter III.27. Obviously, if a Banach algebra has an identity element then every maximal commutative subalgebra has an identity. What about the converse?

The corresponding question for *-algebras was solved by T. Ogasawara [205] and (27.1) is due to him. Results on the above problem would be of interest.

<u>Problem</u>. Let A be a Banach *-algebra. If every maximal commutative *-subalgebra of A has a bounded approximate identity, does A have an approximate identity (bounded or unbounded)?

We make a few remarks concerning conditions on a Banach algebra which imply that it has an identity or approximate identity.

If A is a semisimple commutative Banach algebra with compact carrier space Φ_A, then A has an identity [29, p. 109]. Since $C_o(S)$ is semisimple and its carrier space is S, this result is a substantial improvement over (12.1).

<u>Problem</u>. If A is a semisimple commutative Banach algebra with σ-compact carrier space, does A have an approximate identity? Can it be taken sequential if one exists?

Let S be a C^*-algebra and let S be its unit sphere. Then A has an identity if and only if S has an extreme point [239, p. 10].

<u>Problem</u>. Find the most general class of Banach algebras for which the preceding result holds. Could it be that it characterizes C^*-algebras?

NOTES AND REMARKS

It should be mentioned that if A is a Banach algebra with identity 1, then 1 is an extreme point of the unit sphere of A. This result is due to S. Kakutani [239, 1.6.6, p. 13]. The converse is false. Indeed, if G is an infinite compact group, then $L^2(G)$, under convolution product, is a Banach algebra without identity, but its unit sphere has many extreme points.

28. D. C. Taylor [261] introduced the notion of well-behaved approximate identity and used it to extend a classical theorem of R. S. Phillips to a C^*-algebra setting. By virtue of this extension he made improvements in the work of several authors. H. S. Collins and R. A. Fontenot began a systematic study of well-behaved approximate identities, among others, in [53]. Examples (28.2), (28.3) and Propositions (28.4) and (28.16) are due to them. Example (28.6) and Theorem (28.8) are due to Taylor [261]. Collins and Fontenot conjectured Theorem (28.9), which characterizes paracompactness in terms of well-behaved approximate identities; this theorem was proved by R. F. Wheeler [283] with the help of a deep set-theoretic lemma of A. Hajnal (see Lemma (28.15)).

Recall that a topological space S is <u>pseudocompact</u> if every continuous real-valued function on S is bounded. It is proved in [53, p. 71] that if S is a locally compact, pseudocompact, Hausdorff space, and $C_o(S)$ has a well-behaved approximate identity, then S is compact.

Closely related to, but distinct from, the well-behaved approximate identities are the so-called β-<u>totally bounded approximate identities</u>. We would be remiss if we did not at least define them and point out some of their properties since we have not discussed them in the text, and they have been the subject of considerable research.

The double centralizer algebra $M(A)$ of a C^*-algebra A (see §29) was studied by R. C. Busby [46] who defined the strict topology β as that topology on $M(A)$ generated by the seminorms $x \to \max\{||xy||, ||yx||\}$ for $x \in M(A)$ and $y \in A$. The algebra A lies in $M(A)$ as a closed two-sided ideal, and $M(A)$ always has an identity. When A is commutative, A is $C_o(S)$ for some locally compact Hausdorff space S, and $M(A)$ is $C_b(S)$, the algebra of all bounded continuous complex functions on S. If A is the algebra of compact operators on a Hilbert space H, then it turns out that $M(A)$ is the algebra $B(H)$ of all bounded linear operators on H. By $M(A)_\beta$, we mean $M(A)$ furnished with the strict topology β.

A locally compact Hausdorff space S will be called <u>sham compact</u> if each σ-compact subset is relatively compact.

Let A be a C^*-algebra. An approximate identity $\{e_\lambda\}_{\lambda\varepsilon\Lambda}$ in A is said to be:

(a) β-<u>totally bounded</u> if $\{e_\lambda\}_{\lambda\varepsilon\Lambda}$ is a totally bounded subset of A in the strict topology;

(b) <u>chain totally bounded</u> if whenever $\{\lambda_n\}$ is an increasing sequence in Λ, then $\{e_{\lambda_n}\}$ is β-totally bounbed.

(c) <u>sham compact</u> if $\{e_\lambda\}_{\lambda\varepsilon\Lambda}$ is canonical and if whenever $\{\lambda_n\}$ is a sequence in Λ, then there is a $\lambda \varepsilon \Lambda$ so that $\lambda > \lambda_n$ for all n.

In [52, p. 160], Collins and Dorroh show that if S is paracompact then $C_o(S)$ has a β-totally bounded approximate identity, and ask, among other things, if the converse is true. In [53] Collins and Fontenot prove the converse by a constructive and somewhat lengthy method. Later Wheeler [283] gave a short, nonconstructive, proof of the converse.

It is easy to give an example of a β-totally bounded approximate identity in $C_o(S)$ that is not canonical (and a fortiori, not well-behaved). Example (28.2) provides us with a well-behaved approximate identity that is not β-totally bounded; indeed, the infinite sequence $\{f(i,1)\}$ is clearly not β-totally bounded. However, it is proved in [53], Proposition 7.4, that if $\{e_\lambda\}$ is a canonical chain totally bounded approximate identity for $C_o(S)$, then $\{e_\lambda\}$ is well-behaved.

When A is a noncommutative C^*-algebra, Collins and Fontenot show that a canonical chain totally bounded approximate identity is "almost" well-behaved (see [53, p. 76]), and they conjecture that D. C. Taylor's theorems in [261] remain valid under this (weaker) hypothesis.

In Theorem 5.6 of [53] it is shown that S is a sham compact space if and only if $C_o(S)$ has a sham compact approximate identity.

Finally, we mention that if A is a C^*-algebra with a well-behaved approximate identity, then $M(A)_\beta$ is a strong Mackey space and its dual $M(A)_\beta^*$ is weakly sequentially complete (see [261, p. 483]). For more information on well-behaved and β-totally bounded approximate identities see [52], [53], [260], [261], [283] and their bibliographies.

29. Quasicentral approximate identities were discovered independently and nearly simultaneously by W. Arveson [16] and C. A. Akemann and G. K. Pedersen [5] (the concept appeared in "embryonic form" a little earlier, see [5]). Let A be a C^*-algebra and I a closed two-sided ideal of A. An increasing net $\{e_\lambda\}_{\lambda\in\Lambda}$ of positive elements in I, $||e_\lambda|| \leq 1$ for

all $\lambda \in \Lambda$, which is an approximate identity for I and which satisfies $||e_\lambda x - xe_\lambda|| \to 0$ for every $x \in A$ is called a quasicentral approximate identity for A. The extended definition to more general algebras found in (29.1) is due to A. M. Sinclair [251]. Theorems (29.2) and (29.3) are essentially [16, Theorem 1], and [5, Theorem 3.2]. The proof of Theorem (29.3) given here utilizes a result from D. C. Taylor [260, Theorem 2.1], in conjunction with Theorem (29.2). Theorems (29.5) and (29.6) are due to Sinclair [251].

Problem. Is the hypothesis of separability necessary in Theorem (29.5)?

Both Arveson and Akemann-Pedersen introduced quasicentral approximate identities to study nontrivial problems involving "perturbations" of elements (or representations) of C^*-algebras. In the context of his paper Arveson [16] observes that quasicentral approximate identities are the common thread which can be used to tie together D. Voiculescu's results on (classes of) extensions of the compact operators by a given separable C^*-algebra, Bunce and Salinas' work on matrix ranges of operators, Choi and Effros' work on completely positive liftings of C^*-algebras, as well as several other related problems. Not only do quasicentral approximate identities tie these results together into a neat package, but they also simplify many of the arguments. For further discussion of these interesting results we refer the reader to [5], [16], and a recent preprint by D. W. Hadwin, Approximate equivalence and completely positive maps which, among many other things, proves a nonseparable version of Voiculescu's theorem for arbitrary unital representations of arbitrary C^*-algebras.

We make an additional remark which is not directly related to the material of this section. R. J. Archbold [14] defined a C^*-algebra to be <u>quasicentral</u> if no primitive ideal of A contains the center $Z(A)$ of A. Archbold proved that in order for A to be quasicentral it is necessary and sufficient that A have an approximate identity each element of which belongs to $Z(A)$. This result of Archbold was extended by S. Takahashi [257] to a more general context.

30. The algebra $B'(H)$ of compact operators on a Hilbert space H is well understood [242]. Many results can be extended to the algebra $B'(X)$ of compact operators on a classical Banach space. However little is known for general Banach spaces. Therefore it is not surprising that even the following problem is still unsolved.

<u>Problem</u>. Let $B'(X)$ be the algebra of compact operators on a Banach space X. Does $B'(X)$ have an approximate identity? Is $B'(X)$ factorable?

The few results presented in this section are well known from the study of bases in Banach spaces. For the general theory of bases we refer to the books of J. T. Marti [186] and I. Singer [252]; for a discussion of the approximation property see A. Grothendieck's monograph [119]; for a careful treatment of the theory of classical Banach spaces and many interesting open questions see the recent book [180] by J. Lindenstrauss and L. Tzafriri. The developments in basis theory up to 1972 have been surveyed by C. W. McArthur [190]. Much work has been done since then and we refer to [180] for a description of it.

Example (30.6) is due to B. E. Johnson [139], Lemma 6.2.

Approximate identities for the algebra of finite rank operators are discussed by E. Berkson and H. Porta [24].

31. Many Banach algebras in analysis are ideals in another Banach algebra. The _Segal algebras_ studied by H. Reiter [218], [219], [220] are ideals in the group algebra $L^1(G)$ of a locally compact group G. J. Cigler [48] studied more general "normed ideals" in $L^1(G)$. The C_p algebras [77], pp. 1088-1119, of operators on a Hilbert space H are ideals in the algebra of compact operators on H; C_1 is the ideal of _trace class_ operators on H and C_2 is the ideal of _Hilbert-Schmidt_ operators on H.

The notion of an _abstract Segal algebra_ was introduced by J. T. Burnham [35]. The general theory of abstract Segal algebras was greatly simplified by the work of B. A. Barnes [22]. Closed ideals in Segal algebras have been studied by many authors, e.g. [22], [35], [88], [89], [90], [91], [177], [219], [220], [278].

The method of linear functionals has been employed by H. Reiter [220], §11, Proposition 12; §16, Theorem 1; M. Leinert [177]; and B. E. Johnson [141].

32. The closedness of the sum of a left and a right ideal in a C^*-algebra was first noted by F. Combes [55], Proposition 6.2; he attributes the proof of this result to F. Perdrizet. W. Rudin [237], unaware of Combes' paper, formulated the theorem concerning sums of subspaces of a Banach space; his proof is essentially the same as in [55].

This result enabled him to give a beautiful short proof of the theorem of Sarason, which states that $H^{\infty} + C$ is a closed subalgebra of L^{∞} on the unit circle. We remark that D. A. Stegenga [253] proved a converse of (32.1). For further applications and counterexamples see [237].

Problem. Does the conclusion of Theorem (32.1) remain true for incomplete normed modules X?

We know that if I is a closed ideal in a normed algebra A such that A/I has a left approximate identity and I has a bounded left approximate identity, then A has a left approximate identity (see (7.1)). The method of linear functionals and (32.1) can be used to give an alternate proof of this. Possibly these techniques can be used to solve other problems involving approximate identities.

33. The notion of weak approximate identity was introduced for the study of the second dual A^{**} of a Banach algebra A, see, for example, [49], [125], [188], [189]. Proposition (33.2) can be found in [29, p. 58]. If A is a commutative Banach algebra with carrier space Φ_A, a weak approximate identity is defined by C. A. Jones and C. D. Lahr [146] to be a net $\{e_\lambda\}_{\lambda \varepsilon \Lambda}$ in A such that $\phi(e_\lambda x) \to \phi(x)$ for all $a \varepsilon A$ and $\phi \varepsilon \Phi_A$. They show by a counterexample that, with this definition of weak approximate identity, Proposition (33.2) is false. Their counterexample is a convolution semigroup algebra with a weak approximate identity (as defined above) bounded by one, but with no approximate identity, bounded or unbounded.

The idea of defining a multiplication in the second dual A^{**} of a

Banach algebra A which makes A^{**} into a Banach algebra and extends the product on A is due to R. Arens [15]. The first systematic study of this notion was undertaken by P. Civin and B. Yood [49], and Theorem (33.3) is due to them. J. Hennefeld [125], [126], [127] has studied the problem of when the Arens product is regular. E. A. McCharen [191] has utilized the Arens products to define a locally convex topology on a Banach algebra A which has several interesting properties.

The Arens product in locally convex topological algebras has been studied by G. F. Bombal [26], S. L. Gulick [121], and J. E. Simpson [248].

For further information and additional references to Arens products see [25], [29], [124], [187], [188], [189], [192], [286], [287].

34. The results in this section are taken from J. W. Baker and J. S. Pym [21]. The last statement of Corollary (34.3) which states that $||a|| \leq \beta|a|_\sigma$ for all $a \in A$ is true for normed algebras in general, i.e., the approximate identity can be deleted from the hypothesis (see R. A. Hirschfeld and W. Zelazko, <u>On spectral norm Banach algebras</u>, Bull. Acad. Polon. Sci. Sér. Sci. Math. Astronom. Phys. 16 (1968), 195-199). Baker and Pym give an example showing (34.1) may fail if the scalars are not complex, and reformulate it for real Banach algebras.

35. Both Theorem (35.1) and Example (35.2) were discovered by C. A. Akemann [4].

36. Approximate diagonals were introduced by B. E. Johnson [140]

APPENDIX

We see that (1) follows immediately from Lemma (A.4) since,

in case (2),

$$||R(r_1'\cdots r_m' r_1\cdots r_n)|| = ||r_1'|| = ||r_1'\cdots r_m'|| \le ||r_1'\cdots r_m'||\cdot||r_1\cdots r_n||;$$

in case (3),

$$||R(r_1'\cdots r_m' r_1\cdots r_n)|| = ||r_1\cdots r_n|| \le ||r_1'\cdots r_m'||\cdot||r_1\cdots r_n||;$$

in case (4),

$$||R(r_1'\cdots r_m' r_1\cdots r_n)|| = 1 \le ||r_1'\cdots r_m'||\cdot||r_1\cdots r_n||.$$

To prove Lemma (A.4) we need the following preliminary result.

(A.5) <u>Lemma</u>. Let r_λ denote either s_λ or t_λ, and $|r_\lambda| = \lambda$. Then either

$$R(r_\lambda r_1\cdots r_n) = r_\lambda r_k\cdots r_n \tag{5}$$

for some $1 \le k \le n+1$, or

$$R(r_\lambda r_1\cdots r_n) = r_1\cdots r_n, \tag{6}$$

in which case $|r_1| \le \lambda$, or

$$R(r_\lambda r_1\cdots r_n) = r_k\cdots r_n \tag{7}$$

for some $1 < k \le n$ with $r_k = s_\gamma$ for some γ.

<u>Proof that Lemma (A.5) implies Lemma (A.4)</u>: Our proof is by induction on m. For $m = 1$, (A.4) follows directly from (A.5), with cases (2), (3), (4) coming from (5), (6), (7), respectively.

For $m > 1$, assuming the result holds for $m-1$, we have

$$R(r_1'\cdots r_m' r_1\cdots r_n) = R(r_1' R(r_2'\cdots r_m' r_1\cdots r_n)),$$

and, by the induction hypothesis, there are three cases to consider.

First, if we have

$$R(r'_2 \cdots r'_m r_1 \cdots r_n) = r'_2 \cdots r'_q r_k \cdots r_n$$

for some $2 \leq q \leq m$, $1 \leq k \leq n+1$, then

$$R(r'_1 \cdots r'_m r_1 \cdots r_n) = r'_1 \cdots r'_q r_k \cdots r_n.$$

Secondly, if

$$R(r'_2 \cdots r'_m r_1 \cdots r_n) = r_1 \cdots r_n,$$

then

$$R(r'_1 \cdots r'_m r_1 \cdots r_n) = R(r'_1 \cdots r_n)$$

and the result follows from the m = 1 case.

Thirdly, if

$$R(r'_2 \cdots r'_m r_1 \cdots r_n) = r_k \cdots r_n$$

with $1 < k \leq n$, then $r_k = s_\gamma$ for some γ and

$$R(r'_1 \cdots r'_m r_1 \cdots r_n) = R(r'_1 r_k \cdots r_n)$$

and, when we apply Lemma (A.5) to $R(r'_1 r_k \cdots r_n)$, (5) gives (2) with $q = 1$, (6) gives (4) since $r_k = s_\gamma$, and (7) gives (4).

<u>Proof of Lemma</u> (A.5). Our proof is by induction on n. For n = 1, inspection of the defining relations shows that $R(r_\lambda r_1) = r_\lambda r_1$ or r_1. Hence we have (5) for k = 1, (5) for s = 2, or (6). From the defining relations again, we see that (6) occurs only if $|r_1| \leq \lambda$.

For n > 1, assuming the result for n-1, we argue that

APPENDIX

$$R(r_\lambda r_1 \cdots r_n) = R(R(r_\lambda r_1) r_2 \cdots r_n).$$

If $R(r_\lambda r_1) = r_\lambda r_1$, no reduction occurs and we have (5) with $k = 1$. If $R(r_\lambda r_1) = r_1$, (in which case $|r_1| \leq \lambda$), no further reduction occurs since $r_1 \cdots r_n$ is assumed reduced, so we have (6). Suppose $R(r_\lambda r_1) = r_\lambda$, which, by the defining relations, occurs only if $|r_1| \geq \lambda$. Then

$$R(r_\lambda r_1 \cdots r_n) = R(r_\lambda r_2 \cdots r_n)$$

and, by the induction hypothesis, there are three cases to consider. First, if

$$R(r_\lambda r_2 \cdots r_n) = r_\lambda r_k \cdots r_n$$

for some $2 \leq k \leq n+1$, then we have (5). Secondly, if

$$R(r_\lambda r_2 \cdots r_n) = r_2 \cdots r_n$$

with $|r_2| \leq \lambda$, then we have (7) with $k = 2$. However, since this gives $|r_2| \leq \lambda \leq |r_1|$, the defining relations show that the product $r_1 r_2$ reduces (contrary to hypothesis) unless $r_1 = t_\alpha$ and $r_2 = s_\gamma$ for some α, γ. Hence we have (7) with $r_k = s_\gamma$, as desired. Finally, if

$$R(r_\lambda r_2 \cdots r_n) = r_k \cdots r_n$$

for some $2 < k \leq n$ with $r_k = s_\gamma$ for some γ, then we have (7) for this k. This completes the proof of Lemma (A.5) and also Lemma (A.3). □

We next verify the statements made about the function $\underline{H}'$ defined on p. 31 in the proof of (4.1). That $\underline{H}'$ is monotonically increasing follows directly from part (b) of the following lemma and the fact that

$(H + \varepsilon)^n$ and nH^n are increasing functions of n.

(A.6) Lemma. With N chosen as on p. 31, we have:

(a) $N \geq 3$;

(b) $(H + \varepsilon)^{N-1} \leq (N - 1)H^{N-1}$;

(c) $N > H\varepsilon^{-1}$.

Proof. (a) Since $\varepsilon < (\sqrt{2} - 1)H$, we have $(H + \varepsilon)^2 < 2H^2$, and so $N \neq 2$.

(b) Since N is defined to be minimal subject to two conditions, the number $N-1$ must fail to satisfy one of these. But part (a) shows that $N-1 \geq 2$, so we must have $(H + \varepsilon)^{N-1} \not> (N-1)H^{N-1}$.

(c) We know that $(1 + x^{-1})^x < e$ $(x > 0)$, and that $(1 + x^{-1})^y$ is an increasing function of y. Therefore, if $N \leq H\varepsilon^{-1}$, then

$$\begin{aligned}(1 + \tfrac{\varepsilon}{H})^N &< e \\ &< N \quad \text{by part (a).}\end{aligned}$$

This contradicts the definition of N, so $N > H\varepsilon^{-1}$. □

(A.7) Corollary. The function $\underline{H}'$ satisfies $\underline{H}'(m + n) \leq \underline{H}'(m)\underline{H}'(n)$ for all $m, n \in Z^+$.

Proof. Case 1. Suppose $m, n, m+n < N$. Then $\underline{H}'(m + n) = (H + \varepsilon)^{m+n} = \underline{H}'(m)\underline{H}'(n)$.

Case 2. Suppose $m, n < N$, but $m+n \geq N$. We observe that

$$rH^r < (H + \varepsilon)^r \qquad (r \geq N), \tag{8}$$

since this is true for $r = N$, by the definition of N, and, if it is true for r, then it is true for $r+1$, because

$$\left(\frac{H+\varepsilon}{H}\right)^{r+1} > r\left(1+\frac{\varepsilon}{H}\right)$$

$$> r + 1, \text{ since } r \geq N > H/\varepsilon, \text{ by (A.6),(c).}$$

Case 2 now follows from (8) and the argument in Case 1.

<u>Case</u> 3. Suppose $m, n \geq N$. Then

$$\begin{aligned} \underline{H}'(m+n) &= (m+n)H^{m+n} \\ &\leq mnH^{m+n}, \quad \text{since } N \geq 1, \text{ and so } m, n \geq 2, \\ &= \underline{H}'(m)\underline{H}'(n). \end{aligned}$$

<u>Case</u> 4. Suppose $m < N \leq n$, (the case $n < N \leq m$ being similar).

Then

$$\begin{aligned} \frac{H^n(H+\varepsilon)^m}{H^{m+n}} &= (1+\frac{\varepsilon}{H})^m \\ &> (1+\frac{1}{n})^m, \qquad \text{since } n \geq N > H\varepsilon^{-1}, \\ &\geq 1+\frac{m}{n}, \qquad \text{by the Binomial Theorem,} \\ &= \frac{m+n}{n}. \end{aligned}$$

Hence $\underline{H}'(m+n) = (m+n)H^{m+n} \leq (H+\varepsilon)^m \cdot nH^n = \underline{H}'(m)\underline{H}'(n)$. □

We next wish to make a few remarks concerning the proof of Lemma (28.11). It is non-trivial to verify that $\{f_\lambda \circ \psi\}_{\lambda \varepsilon \Lambda}$ is a w.a.i. for $C_o(Y)$ if $\{f_\lambda\}_{\lambda \varepsilon \Lambda}$ is not bounded. It <u>is</u> clear that $f_\lambda \to 1$ locally uniformly in X and so $f_\lambda \circ \psi \to 1$ locally uniformly in Y, whence $f_\lambda \circ \psi . F \to F$ in $C_o(Y)$ for each function F with compact support. To achieve this for <u>all</u> $F \varepsilon C_o(Y)$ we use the following lemma which was pointed out to us by R. B. Burckel:

(A.8) Lemma. If $F \in C_o(Y)$, then there exists $f \in C_o(X)$ such that $f \circ \psi \geq |F|$.

Proof. We may assume without loss of generality that $||F||_\infty = 1$. Let $K_n = \{y \in Y: 2^{-n} \leq |F(y)| \leq 2^{-n+1}\}$. This is a compact subset of Y and so $L_n = \psi(K_n)$ is a compact subset of X. Let $\phi_n \in C_o(X)$ satisfy $0 \leq \phi_n \leq 1 = \phi_n(L_n)$, and set $f = \Sigma_{n=1}^{\infty} 2^{-n+1}\phi_n$. If $F(y) \neq 0$, there exists a positive integer n so that $y \in K_n$. Therefore $\psi(y) \in L_n$ and so $\phi_n(\psi(y)) = 1$. Hence, $f(\psi(y)) \geq 2^{-n+1}\phi_n(\psi(y)) = 2^{-n+1} \geq |F(y)|$. □

Utilizing Lemma (A.8), (28.11) follows from:

$$|f_\lambda \circ \psi \cdot F - F| = |f_\lambda \circ \psi - 1||F| \leq |f_\lambda \circ \psi - 1||f \circ \psi|$$
$$= |f_\lambda \cdot f - f| \circ \psi;$$

hence $||f_\lambda \circ \psi \cdot F - F||_\infty \leq ||f_\lambda \cdot f - f||_\infty \to 0$. □

Lemma (A.8) is also useful to clarify a point in the proof of Proposition (28.16). Indeed, to move from $\{e_\lambda \otimes f_\alpha\}$ being an approximate identity for the dense subalgebra generated by $\{f \otimes g\}$ to its being an approximate identity (sup-norm) for all of $C_o(X \times Y)$ is unclear in the absence of a boundedness hypothesis on $\{e_\lambda\}$ and $\{f_\alpha\}$. However, given $F \in C_o(X \times Y)$, there exists (A.8) a functions $f \in C_o(X)$ and $g \in C_o(Y)$ such that $f \otimes g \geq |F|$. Therefore $|e_\lambda \otimes f_\alpha \cdot F - F| = |e_\lambda \otimes f_\alpha - 1||F| \leq |e_\lambda \otimes f_\alpha - 1||f \otimes g| = |e_\lambda \otimes f_\alpha \cdot f \otimes g - f \otimes g|$.

Our final result in this appendix concerns Theorem (35.1). We wish to observe that when the C^*-algebra A is commutative, then the inequality $||b|| \leq ||a_o|| + \varepsilon$ can be replaced by $||b|| = ||a_o||$.

APPENDIX

Indeed, we may view A as $C_o(X)$ for some locally compact Hausdorff space X, with B a C^*-subalgebra of A. Then, if $a \geq 0$ belongs to A, there exists (by (35.1)) $b \in B$ with $b \geq a$. Let $\hat{1}$ denote the function identically equal to 1, and set $b_o = b \wedge ||a||\cdot\hat{1}$ (where we are viewing them as functions.) Clearly, $b_o(t) = b(t)$ or $||a||$ for each $t \in X$, and hence $b_o(t) \geq a(t)$, i.e., $b_o \geq a$. Also $b_o(t) \leq ||a||$, so that $||b_o|| \leq ||a||$; thus $||b_o|| = ||a||$. To see that $b_o \in B$, simply note that b_o clearly belongs to the algebra of bounded continuous functions generated by B and $\hat{1}$. Since b vanishes at infinity, so does b_o, and hence $b_o \in B$.

BIBLIOGRAPHY

1. J. F. Aarnes and R. V. Kadison, Pure states and approximate identities. Proc. Amer. Math. Soc. 21(1969), 749-752. MR 39 #1980. Z. 177, 413.

2. C. A. Akemann, Interpolation in W*-algebras. Duke Math. J. 35(1968), 525-533. MR 37 #4626. Z. 172, 412.

3. C. A. Akemann, Invariant subspaces of C(G). Pacific J. Math. 27(1968), 421-424. MR 38 #5966. Z. 167, 144.

4. C. A. Akemann, Approximate units and maximal abelian C*-subalgebras. Pacific J. Math. 33(1970), 543-550. MR 41 #9001. Z. 184, 169.

5. C. A. Akemann and G. K. Pedersen, Ideal perturbations of elements in C*-algebras. Math. Scand. 41(1977), 117-139.

6. C. A. Akemann and M. Rosenfeld, Maximal one-sided ideals in operator algebras. Amer. J. Math. 94(1972), 723-728. MR 47 #7447.

7. G. R. Allan and A. M. Sinclair, Power factorization in Banach algebras with bounded approximate identity. Studia Math. 56(1976), 31-38. MR 53 #14130. Z. 338.46044.

8. M. Altman, Factorisation dans les algèbres de Banach. C. R. Acad. Sci. Paris 272(1971), 1388-1389. MR 44 #2038. Z. 232.46045.

9. M. Altman, Infinite products and factorization in Banach algebras. Bollettino U. M. I. 5(1972), 217-229. MR 50 #14233. Z. 252.46043.

10. M. Altman, Contracteurs dans les algèbres de Banach. C. R. Acad. Sci. Paris 274(1972), 399-400. MR 45 #2473. Z. 232.46044.

11. M. Altman, Contractors, approximate identities and factorization in Banach algebras. Pacific J. Math. 48(1973), 323-334. MR 52 #11593. Z. 281.46044.

12. M. Altman, A generalization and the converse of Cohen's factorization theorem. Duke Math. J 42(1975), 105-110. MR 52 #15010. Z. 281.46049.

13. M. Altman, Contractors and contractor directions, theory and applications. Lecture notes in pure and applied math., 32. Marcel-Dekker. New York. 1977.

14. R. J. Archbold, Density theorems for the centre of a C*-algebra. J. London Math. Soc. (2) 10(1975), 189-197. MR 52 #1331. Z. 299.46057.

BIBLIOGRAPHY

15. R. Arens, The adjoint of a bilinear operation. Proc. Amer. Math. Soc. 2(1951), 839-848. MR 13, 659. Z. 44, 326.

16. W. Arveson, Notes on extensions of C^*-algebras. Duke Math. J., 44(1977), 329-355. MR 55 #11056. Z. 368.46052.

17. G. F. Bachelis, Homomorphisms of annihilator Banach algebras. II. Pacific J. Math. 30(1969), 283-291. MR 40 #7798. Z. 179, 461.

18. G. F. Bachelis, Homomorphisms of Banach algebras with minimal ideals. Pacific J. Math. 41(1972), 307-311. MR 47 #5597.

19. G. F. Bachelis, W. A. Parker and K. A. Ross, Local units in $L^1(G)$. Proc. Amer. Math Soc. 31(1972), 312-313. MR 44 #5794. Z. 252.43013.

20. W. G. Bade and P. C. Curtis, The continuity of derivations of Banach algebras. J. Functional Analysis 16(1974), 372-387. MR 50 #10820. Z. 296.46049.

21. J. W. Baker and J. S. Pym, A remark on continuous bilinear mappings. Proc. Edinburgh Math. Soc. 17(1971), 245-248. MR 43 #2429. Z. 217, 448.

22. B. A. Barnes, Banach algebras which are ideals in a Banach algebra. Pacific J. Math. 38(1971), 1-7; correction, ibid. 39(1971), 828. MR 46 #9738. Z. 226.46054.

23. J. Benedetto, Spectral synthesis, Academic Press. New York, 1975. Z. 314.43011.

24. E. Berkson and H. Porta, Representations of B(X). J. Functional Analysis 3(1969), 1-34. MR 38 #3728. Z. 169, 174.

25. F. T. Birtel, On a commutative extension of a Banach algebra. Proc. Amer. Math. Soc. 13(1962), 815-822. MR 31 #624. Z. 112, 344.

26. G. F. Bombal, The Aren's product in locally convex algebras (Spanish). Rev. Real Acad. Ci. Exact. Fiś. Natur. Madrid 70(1976), 337-346. MR 55 #1066. Z. 338.46039.

27. F. F. Bonsall and J. Duncan, Numerical ranges of operators on normed spaces and of elements of normed algebras. Cambridge Univ. Press. 1971. MR 44 #5779. Z. 207, 448.

28. F. F. Bonsall and J. Duncan, Numerical ranges. II. Cambridge Univ. Press. 1973. Z. 262.47001.

29. F. F. Bonsall and J. Duncan, Complete normed algebras. Springer-Verlag, Berlin. 1973. MR 54 #11013. Z. 271.46039.

BIBLIOGRAPHY

30. N. Bourbaki, Théories spectrales, Chapitres 1, 2. Hermann, Paris, 1967. MR 35 #4725. Z. 152, 326.

31. A. Browder, Introduction to function algebras. W. A. Benjamin, Inc., New York, 1969. MR 39 #7431. Z. 199, 461.

32. R. C. Buck, Operator algebras and dual spaces. Proc. Amer. Math. Soc. 3(1952), 681-687. MR 14, 290. Z. 47, 315.

33. R. C. Buck, Bounded continuous functions on a locally compact space. Mich. Math. J. 5(1958), 95-104. MR 21 #4350. Z. 87, 315.

34. J. T. Burnham, Closed ideals in subalgebras of Banach algebras. II. Notices Amer. Math. Soc. 17(1970), 815. Abstract #70T-B149.

35. J. T. Burnham, Closed ideals in subalgebras of Banach algebras. I. Proc. Amer. Math. Soc. 32(1972), 551-555. MR 45 #4146. Z. 234.46050.

36. J. T. Burnham, Nonfactorization in subsets of the measure algebra. Proc. Amer. Math. Soc. 35(1972), 104-106. MR 45 #7394. Z. 221.43004.

37. J. T. Burnham, Closed ideals in subalgebras of Banach algebras. II. Ditkins condition. Monatsh. Math. 78(1974), 1-3. MR 49 #5845. Z. 275.46039.

38. J. T. Burnham, Segal algebras and dense ideals in Banach algebras. Lecture Notes in Math 399. Springer-Verlag, New York 1974 pp. 33-58. MR 54 #3295. Z. 303.43010.

39. J. T. Burnham, Relative completions of A-Segal algebras. Bull. Amer. Math. Soc. 81(1975), 125-127. MR 50 #14230. Z. 333.46037.

40. J. T. Burnham, The relative completion of an A-Segal algebra is closed. Proc. Amer. Math. Soc. 49(1975), 116-122. MR 50 #14231. Z. 301.46035.

41. J. T. Burnham, Multipliers of commutative A-Segal algebras. Tamkang J. Math. 7(1976), 7-18. MR 55 #980.

42. J. T. Burnham and R. R. Goldberg, Basic properties of Segal algebras. J. Math. Anal. Appl. 42(1973), 323-329. MR 48 #4648. Z. 258.43009.

43. J. T. Burnham and R. R. Goldberg, The convolution theorems of Dieudonné. Acta Sci. Math. (Szeged) 36(1974), 1-3. MR 49 #11149. Z. 295.43003.

44. J. T. Burnham and R. R. Goldberg, Multipliers from $L^1(G)$ to a Segal algebra. Bull. Inst. Math. Acad. Sinica 2(1974), 153-164. MR 51 #1284. Z. 292.43009.

BIBLIOGRAPHY

45. J. T. Burnham and P. S. Muhly, Multipliers of commutative Segal algebras. Tamkang J. Math., 6(1975), 229-238. MR 54 #3299. Z. 338.43006.

46. R. C. Busby, Double centralizers and extensions of C*-algebras. Trans. Amer. Math. Soc. 132(1968), 79-99. MR 37 #770. Z. 165, 155.

47. H. Cartan and R. Godement, Théorie de la dualité et analyse harmonique dans les groupes abéliens localement compacts. Ann. Sci. École Norm. Sup. (3) 64(1947), 79-99. MR 9, 326.

48. J. Cigler, Normed ideals in $L^1(G)$. Nederl. Akad. Wetensch. Proc. Ser. A 72 = Indag. Math. 31(1969), 273-282. MR 40 #3327. Z. 176, 447.

49. P. Civen and B. Yood, The second conjugate space of a Banach algebra as an algebra. Pacific J. Math. 11(1961), 847-870. MR 26 #622. Z. 117, 97.

50. R. Rao Chivukula and R. K. Heckman, Convolutions and factorization theorems. Compositio math 28(1974), 143-157. MR 49 #7701. Z. 284.43003.

51. P. J. Cohen, Factorization in group algebras. Duke Math. J. 26(1959), 199-205. MR 21 #3729. Z. 85, 102.

52. H. S. Collins and J. R. Dorroh, Remarks on certain function spaces. Math. Annalen, 176(1968), 157-168. MR 36 #5694. Z. 159, 184.

53. H. S. Collins and R. A. Fontenot, Approximate identities and the strict topology. Pacific J. Math. 43(1972), 63-79. MR 47 #2378. Z. 257.46049.

54. H. S. Collins and W. H. Summers, Some applications of Hewitt's factorization theorem. Proc. Amer. Math. Soc. 21(1969), 727-733. MR 39 #1982. Z. 174, 185.

55. F. Combes, Sur les faces d'une C^*-algèbre. Bull. Sci. Math. 93(1969), 37-62. MR 42 #856. Z. 177, 178.

56. C. V. Comisky, Multipliers of Banach modules. Nederl. Akad. Wetensch. Proc. Ser. A 74 = Indag. Math 33(1971), 32-38. MR 43 #5307. Z. 213,138.

57. I. G. Craw, Factorisation in Fréchet algebras. J. London Math. Soc. 44(1969), 607-611. MR 39 #3311. Z. 175, 142.

58. I. G. Craw, Axiomatic cohomology for Banach modules. Proc. Amer. Math. Soc. 38(1973), 68-74. MR 47 #7433. Z. 238.46051.

59. P. C. Curtis, Jr. and A. Figá-Talamanca, Factorization theorems for Banach algebras. In: Function algebras, edited by F. T. Birtel, Scott, Foresman and Co., Chicago, Ill., 1966, pp. 169-185. MR 34 #3350. Z. 191, 138.

60. P. C. Curtis, Jr. and H. Stetkaer, A factorization theorem for analytic functions operating in a Banach algebra. Pacific J. Math. 37(1971), 337-343. MR 46 #9735. Z. 203, 447.

61. H. G. Dales, A discontinuous homomorphism from C(X). Amer. J. Math. (to appear).

62. H. G. Dales and J. Esterle, Discontinuous homomorphisms from C(X). Bull. Amer. Math. Soc. 83(1977), 257-259.

63. J. W. Davenport, Multipliers on a Banach algebra with a bounded approximate identity. Pacific J. Math. 63(1976), 131-135. MR 54 #931. Z. 328.46050.

64. A. M. Davie, The approximation problem for Banach spaces. Bull. London Math. Soc. 5(1973), 261-266. MR 49 #3499. Z. 267.46013.

65. A. M. Davie, A counterexample on dual Banach algebras. Bull. London Math. Soc. 5(1973), 79-80. MR 49 #3498. Z. 257.46107.

66. A. M. Davie, The Banach approximation problem. J. Approximation theory 13(1975), 392-394. MR 52 #1262. Z. 299.46019.

67. M. M. Day, Normed linear spaces. Springer-Verlag, Berlin, 1958. MR 20 #1187. Z. 82, 106.

68. A. Derighetti, Some results on the Fourier-Stieltjes algebra of a locally compact group. Comment. Math. Helv. 45(1970), 219-228. MR 54 #856. Z. 211, 155.

69. W. Dietrich, On the ideal structure of Banach algebras. Trans. Amer. Math. Soc. 169(1972), 59-74. MR 46 #7905. Z. 257.46062.

70. J. Dixmier, C^*-algebras. North-Holland Mathematical Library. Vol. 15. North-Holland Publishing Co., Amsterdam 1977. MR 39 #7442.

71. P. G. Dixon, Approximate identities in normed algebras. Proc. London Math. Soc. 26(1973), 485-496. MR 47 #9286. Z. 252.46046.

72. P. G. Dixon, Approximate identities in normed algebras. II. J. London Math. Soc. 17(1978), 141-151. Z. 384.46029.

73. R. S. Doran and J. Wichmann, The Gelfand-Naimark theorems in C^*-algebras. L'Enseignement math. 23(1977), 153-180.

BIBLIOGRAPHY

74. J. Dugundji, Topology. Allyn and Bacon, Boston, 1966. MR 33 #1824. Z. 144, 215.

75. D. H. Dunford, Segal algebras and left normed ideals. J. London Math. Soc. (2) 8(1974), 514-516. MR 49 #11148. Z. 302.43009.

76. N. Dunford and J. T. Schwartz, Linear operators. I. J. Wiley, New York, 1958. MR 22 #8302. Z. 84, 104.

77. N. Dunford and J. T. Schwartz, Linear operators. II. J. Wiley, New York, 1963. MR 32 #6181. Z. 128, 348.

78. N. Dunford and J. T. Schwartz, Linear operators. III. J. Wiley, New York, 1972. MR 54 #1009. Z. 243.47001.

79. R. E. Edwards, Approximation by convolutions. Pacific J. Math. 15(1965), 85-95. MR 33 #6296. Z. 141, 125.

80. R. E. Edwards, Fourier Series. I. Holt, Rinehart and Winston, Inc. New York, 1967. MR 35 #7062. Z. 152, 259.

81. R. E. Edwards, Fourier Series. II. Holt, Rinehart and Winston, Inc. New York, 1967. MR 36 #5588. Z. 189, 66.

82. R. E. Edwards and E. Hewitt, Pointwise limits for sequences of convolution operators. Acta Math. 113(1965), 180-218. MR 31 #1522. Z. 161, 111.

83. G. A. Elliott, A weakening of the axioms for a C^*-algebra. Math. Ann. 189(1970), 257-260. MR 43 #2521. Z. 194, 156; 199, 197.

84. P. Enflo, A counterexample to the approximation problem in Banach spaces. Acta Math. 130(1973), 309-317. Z. 267.46012.

85. M. S. Espelie, Multiplicative and extreme positive operators, Pacific J. Math. 48(1973), 57-66. MR 49 #1212. Z. 247.47037.

86. J. Esterle, Injection de semi-groupes divisibles dans des algèbres de convolution et construction d'homomorphismes discontinus de C(K), Proc. London Math. Soc. (3) 36(1978), 59-85.

87. P. Eymard, L'algèbre de Fourier d'un groupe localement compact. Bull. Soc. Math. France 92(1964), 181-236. MR 37 #4208. Z. 169, 464.

88. H. G. Feichtinger, Zur Idealtheorie von Segal-Algebren. manuscripta math. 10(1973), 307-312. MR 48 #2769. Z. 264.43008.

89. H. G. Feichtinger, Multipliers of Banach spaces of functions on groups. Math. Z. 152(1977), 47-58. MR 55 #982. Z. 324.43005.

BIBLIOGRAPHY

90. H. G. Feichtinger, Results on Banach ideals and spaces of multipliers. Math. Scand. 41(1977), 315-324.

91. H. G. Feichtinger, On a class of convolution algebras of functions. Ann. Inst. Fourier, Grenoble 27(1977), 135-162. Z. 316.43004.

92. H. G. Feichtinger, Multipliers from $L^1(G)$ to a homogeneous Banach space. J. Math. Anal. Appl. 61(1977), 341-356. MR 56 #16257.

93. H. G. Feichtinger, C. C. Graham and E. H. Lakien, Partial converses to Cohen's factorization theorem with application to Segal algebras (preprint).

94. J. M. G. Fell, An extension of Mackey's method to Banach *-algebraic bundles. Memoirs Amer. Math. Soc. 90(1969). MR 41 #4255. Z. 194, 443.

95. J. M. G. Fell, Induced representations and Banach *-algebraic bundles. Lecture Notes in Math. 582. Springer-Verlag. Berlin-Heidelberg-New York. 1977.

96. T. Figiel, Factorization of compact operators and applications to the approximation problem. Studia Math. 45(1973), 191-210. MR 49 #1070. Z. 257.47017.

97. T. Figiel and W. B. Johnson, The approximation property does not imply the bounded approximation property. Proc. Amer. Math. Soc. 41(1973), 197-200. MR 49 #5782. Z. 289.46015.

98. S. Friedberg, Closed subalgebras of group algebras. Trans. Amer. Math. Soc. 147(1970), 117-125. MR 40 #7730. Z. 193, 104.

99. B. R. Gelbaum, Note on the tensor product of Banach algebras. Proc. Amer. Math. Soc. 12(1961), 750-757. MR 24 #A1039. Z. 124, 324.

100. B. R. Gelbaum, Tensor products and related questions. Trans. Amer. Math. Soc. 103(1962), 525-548. MR 25 #2406. Z. 111, 120.

101. B. R. Gelbaum, Tensor products over Banach algebras. Trans. Amer. Math. Soc. 118(1965), 131-149. MR 31 #2629. Z. 135, 358.

102. B. R. Gelbaum, Tensor products of Banach algebras. II. Proc. Amer. Math. Soc. 25(1970), 470-474. MR 41 #4238. Z. 196, 148.

103. J. Gil de Lamadrid, Uniform cross norms and tensor products of Banach algebras. Duke Math. J. 32(1965), 359-368. MR 32 #8125. Z. 135, 360.

104. J. Gil de Lamadrid, Measures and tensors. Trans. Amer. Math. Soc. 114(1965), 98-121. MR 31 #3851.

BIBLIOGRAPHY

105. L. Gillman and M. Jerison, Rings of continuous functions. D. Van Nostrand, Princeton, 1960. MR 22 #6994. Z. 93, 300.

106. R. R. Goldberg, Recent results on Segal algebras. Functional Analysis Appl., internat. conf., Madras 1973, Lecture Notes Math. 399, 220-229. 1974. MR 53 #8783. Z. 303.43009.

107. R. R. Goldberg and S. E. Seltzer, Uniformly concentrated sequences and multipliers of Segal algebras. J. Math. Anal. Appl. 59(1977), 488-497.

108. E. A. Gorin, Spectral stability of some Banach algebras. Functional Analysis Appl. 8(1974) 158-159. MR 50 #14246. Z. 298.46058.

109. C. C. Graham, $M_o(G)$ is not a prime L-ideal of measures. Proc. Amer. Math. Soc. 27(1971), 551-562. MR 43 #3739. Z. 208, 378.

110. M. D. Green, Maximal one-sided ideals in Banach algebras. Math. Proc. Cambridge Philos. Soc. 80(1976), 109-111. MR 53 #14129. Z. 353.46039.

111. W. A. Green, Generalizations and functorial aspects of H^*-algebras. Dissertation, Tulane University, 1973.

112. F. P. Greenleaf, Characterization of group algebras in terms of their translation operators. Pacific J. Math. 18(1966), 243-276. MR 34 #626. Z. 156, 144.

113. F. P. Greenleaf, Invariant means on topological groups. Van Nostrand, New York, 1969. MR 40 #4776. Z. 174, 190.

114. S. Grosser, R. Mosak and M. Moskowitz, Duality and harmonic analysis on central topological groups. I. Indag. Math. 35(1973), 65-77. MR 49 #5225a. Z. 267.22008.

115. S. Grosser, R. Mosak and M. Moskowitz, Duality and harmonic analysis on central topological groups. II. Indag. Math. 35(1973), 78-91. MR 49 #5225b. Z. 267.22008.

116. S. Grosser, R. Mosak and M. Moskowitz, Correction to "Duality and harmonic analysis on central topological groups. I." Indag. Math. 35(1975), 375. MR 49 #5225c. Z. 269.22008.

117. S. Grosser and M. Moskowitz, Harmonic analysis on central topological groups. Trans. Amer. Math. Soc. 156(1971), 419-454. MR 43 #2165. Z. 222.43010.

118. S. Grosser and M. Moskowitz, Compactness conditions in topological groups. J. reine angew. Math. 246(1971), 1-40. MR 44 #1766. Z. 219, 145.

BIBLIOGRAPHY

119. A. Grothendieck, Produits tensoriels topologiques et espaces nucléaires. Memoirs Amer. Math. Soc. 16(1955). MR 17, 763. Z. 64, 355.

120. A. Guichardet, Tensor products of C^*-algebras. I. II. Lecture Notes, Series 12 and 13, Aarhus Universitet, 1969.

121. S. L. Gulick, The bidual of a locally multiplicatively-convex algebra. Pacific J. Math. 17(1966), 71-96. MR 34 #627. Z. 137, 101.

122. S. L. Gulick, T. S. Liu and A. C. M. van Rooij, Group algebra modules. I. II. Canadian J. Math. 19(1967), 133-173. MR 36 #5712, #5713. Z. 148, 120.

123. S. L. Gulick, T. S. Liu and A. C. M. van Rooij, Group algebra modules. III. IV. Trans. Amer. Math. Soc. 152(1970), 561-596. MR 42 #5064. Z. 206, 427.

124. H. Helson, On the ideal structure of group algebras. Ark. Mat. 2(1952), 83-86. MR 14, 246.

125. J. Hennefeld, A note on the Arens products. Pacific J. Math. 26(1968), 115-119. MR 37 #6755. Z. 185, 210.

126. J. Hennefeld, The Arens products and an imbedding theorem. Pacific J. Math. 29(1969), 551-563. MR 40 #4774. Z. 182, 169.

127. J. Hennefeld, Finding a maximal subalgebra on which the two Arens products agree. Pacific J. Math. 59(1975), 93-98. MR 52 #15012. Z. 328.46052.

128. J. N. Henry and D. C. Taylor, The $\overline{\beta}$ topology for W^*-algebras. Pacific J. Math. 60(1975), 123-139. MR 52 #11614. Z. 276.46032.

129. E. Hewitt, The ranges of certain convolution operators. Math. Scand. 15(1964), 147-155. MR 32 #4471. Z. 135, 360.

130. E. Hewitt and K. A. Ross, Abstract harmonic analysis. I. Springer-Verlag, Berlin, 1963. MR 28 #158. Z. 115, 106.

131. E. Hewitt and K. A. Ross, Abstract harmonic analysis. II. Springer-Verlag, New York, 1970. MR 41 #7378. Z. 213, 401.

132. E. Hille and R. S. Phillips, Functional analysis and semigroups. Amer. Math. Soc. Colloquium Publications, Vol. XXXI. Revised Edition. Amer. Math. Soc., Providence, R. I. 1957. MR 19, 664. Z. 78, 100.

133. J. R. Holub, Bounded approximate identities and tensor products. Bull. Austral. Math. Soc. 7(1972), 443-445. MR 48 #6964. Z. 237.46055.

BIBLIOGRAPHY

134. A. Hulanicki and T. Pytlik, On commutative approximate identities and cyclic vectors of induced representations. Studia Math. 48(1973), 189-199. MR 49 #3024. Z. 267.43011.

135. B. E. Johnson, An introduction to the theory of centralizers. Proc. London Math. Soc. 14(1964), 299-320. MR 28 #2450. Z. 143, 361.

136. B. E. Johnson, Centralisers on certain topological algebras. J. London Math. Soc. 39(1964), 603-614. MR 29 #5115. Z. 124, 69.

137. B. E. Johnson, Continuity of centralisers on Banach algebras. J. London Math. Soc. 41(1966), 639-640. MR 34 #629. Z. 143, 361.

138. B. E. Johnson, The uniqueness of the (complete) norm topology. Bull. Amer. Math. Soc. 73(1967), 537-539. MR 35 #2142. Z. 172, 410.

139. B. E. Johnson, Cohomology in Banach algebras. Memoirs Amer. Math. Soc. 127(1972). MR 51 #11130. Z. 256.18014.

140. B. E. Johnson, Approximate diagonals and cohomology of certain annihilator Banach algebras. Amer. J. Math. 94(1972), 685-698. MR 47 #5598. Z. 246.46040.

141. B. E. Johnson, Some examples in harmonic analysis. Studia Math. 48(1973), 181-188. MR 49 #3456. Z. 237.43002.

142. B. E. Johnson, Banach algebras: Introductory course; Introduction to cohomology in Banach algebras. In: Algebras in Analysis, edited by J. H. Williamson. pp. 63-100. Academic Press, New York, 1975. MR 54 #5835.

143. D. L. Johnson, A short proof of the uniqueness of Haar measure. Proc. Amer. Math. Soc. 55(1976), 250-251. MR 53 #774. Z. 324.43001.

144. W. B. Johnson, A complementary universal conjugate Banach space and its relation to the approximation problem. Israel J. Math. 13(1972), 301-310. MR 48 #4700. Z. 252.46024.

145. W. B. Johnson, H. P. Rosenthal and M. Zippin, On bases, finite dimensional decompositions and weaker structures in Banach spaces. Israel J. Math. 9(1971), 488-506. MR 43 #6702. Z. 217, 161.

146. C. A. Jones and C. D. Lahr, Weak and norm approximate identities are different. Pacific J. Math. 72(1977), 99-104. MR 56 #6282.

147. Y. Katznelson, An introduction to harmonic analysis. John Wiley & Sons, Inc., New York, 1968. MR 40 #1734. Z. 169, 179.

BIBLIOGRAPHY

148. C. N. Kellogg, Centralizers and H^*-algebras. Pacific J. Math. 17(1966), 121-129. MR 33 #1749. Z. 139, 76.

149. A. N. Kočubei, The best approximation in normed molules (Russian). Ukrain. Mat. Z. 25(1973), 103-106. MR 47 #829. Z. 262.41035.

150. P. Koosis, Sur un théorème de Paul Cohen. C. R. Acad. Sci. Paris 259(1964), 1380-1382. MR 30 #2295. Z. 175, 142.

151. E. Kotzman and H. Rindler, Central approximate units in a certain ideal of $L^1(G)$. Proc. Amer. Math. Soc. 57(1976), 155-158. MR 53 #8784. Z. 304.43007.

152. E. Kotzman and H. Rindler, Segal algebras on nonabelian groups. Trans. Amer. Math. Soc. 237(1978), 271-281. Z. 363.43004.

153. J. Křížková and P. Vrbová, A remark on a factorization theorem. Comment. Math. Univ. Carolinae 15(1974), 611-614. MR 50 #14234. Z. 329.46055.

154. H. E. Krogstad, On Banach modules with weakly compact action. Institut Mittag-Leffler, 1973.

155. H. E. Krogstad, Multipliers on homogeneous Banach spaces on compact groups. Ark. Mat. 12(1974), 203-212. MR 51 #11014. Z. 297.43005.

156. C. D. Lahr, Approximate identities for convolution measure algebras. Pacific J. Math. 47(1973), 147-159. MR 48 #4739. Z. 272.43001.

157. C. D. Lahr, Multipliers for certain convolution measure algebras. Trans. Amer. Math. Soc. 185(1973), 165-181. MR 48 #11912. Z. 242.43004.

158. C. D. Lahr, Multipliers for ℓ_1-algebras with approximate identities. Proc. Amer. Math. Soc. 42(1974), 501-506. MR 48 #9259. Z. 293.43004.

159. H. C. Lai, Multipliers of a Banach algebra in the second conjugate algebra as an idealizer. Tôhoku Math J. (2)26(1974), 431-452. MR 50 #10817. Z. 293.46037.

160. H. C. Lai, Banach algebras which are ideals in a Banach algebra. Bull. Inst. Math. Acad. Sinica 3(1975), 227-233. MR 53 #1274. Z. 315.46047.

161. E. H. Lakien, Nonfactorization in Segal algebras on compact abelian groups. Dissertation, Northwestern University. 1975.

162. L. J. Lardy, On the identity in a measure algebra. Proc. Amer. Math. Soc. 19(1968), 807-810. MR 37 #3366. Z. 162, 190.

BIBLIOGRAPHY

163. L. J. Lardy and J. A. Lindberg, Jr., On maximal regular ideals and identities in the tensor product of commutative Banach algebras. Canadian J. Math. 21(1969), 639-647. MR 42 #8290. Z. 176, 440.

164. R. Larsen, The multiplier problem. Lecture Notes in Mathematics, Vol. 105. Springer-Verlag, Berlin-New York, 1969. MR 55 #8694. Z. 181, 414.

165. R. Larsen, An introduction to the theory of multipliers. Springer-Verlag, New York-Heidelberg, 1971. MR 55 #8695.

166. R. Larsen, Factorization and multipliers of Segal algebras. Preprint Series, Institute of Math., University of Oslo, #5, 1973.

167. R. Larsen, Tensor product factorization and multipliers. Preprint Series, Institute of Math., University of Oslo, #13, 1974.

168. R. Larsen, The algebras of functions with Fourier transforms in L_p: a survey. Nieuw Arch. Wiskunde, III. Ser 22(1974), 195-240. Z. 291.43004.

169. R. Larsen, The multipliers of $L_1([0,1])$ with order convolution. Publ. Math. Debrecen 23(1976), 239-248. MR 55 #8693.

170. R. Lasser, Primary ideals in centers of group algebras. Math. Ann. 229(1977), 53-58. Z. 337.43004.

171. R. Lasser, Zur Idealtheorie in Gruppenalgebren von $[FIA]_B^-$-Gruppen. Monatsh. Math. 85(1978), 59-79.

172. K. B. Laursen, Tensor products of Banach algebras with involution. Trans. Amer. Math. Soc. 136(1969), 467-487. MR 38 #2610. Z. 174, 186.

173. K. B. Laursen, Ideal structure in generalized group algebras. Pacific J. Math. 30(1969), 155-174. MR 40 #1792. Z. 176, 442.

174. K. B. Laursen, Maximal two-sided ideals in tensor products of Banach algebras. Proc. Amer. Math. Soc. 25(1970), 475-480. MR 44 #802. Z. 195, 138.

175. M. Leinert, A contribution to Segal algebras. manuscripta math. 10(1973), 297-306. MR 48 #2768. Z. 265.46048.

176. M. Leinert, Remarks on Segal algebras. manuscripta math. 16(1975), 1-9. Z. 318.46058.

177. M. Leinert, Some aspects of Segal algebras. Rend. Sem. Mat. Fis. Milano. 45(1975), 65-69. MR 55 #6120.

178. M. Leinert, A commutative Banach algebra which factorizes but has no approximate units. Proc. Amer. Math. Soc. 55(1976), 345-346. MR 53 #1171. Z. 325.46057.

179. M. Leinert, A factorable Banach algebra with inequivalent regular representation norm. Proc. Amer. Math. Soc. 60(1976), 161-162. MR 54 #8151. Z. 325.46058.

180. J. Lindenstrauss and L. Tzafriri, Classical Banach spaces. I. Springer-Verlag, Berlin. 1977.

181. T. S. Liu and A. van Rooij, Sums and intersections of normed linear spaces. Math. Nachrichten 41-42(1969), 29-42. MR 42 #8249. Z. 206, 421.

182. T. S. Liu, A. van Rooij and J. K. Wang, Projections and approximate identities for ideals in group algebras. Trans. Amer. Math. Soc. 175(1973), 469-482. MR 47 #7327. Z. 269.22003.

183. T. S. Liu and J. K. Wang, Sums and intersections of Lebesque spaces. Math. Scand. 23(1968), 241-251. MR 41 #7503. Z. 187, 74.

184. L. H. Loomis, An introduction to abstract harmonic analysis. Van Nostrand, New York, 1953. MR 14, 883. Z. 52, 117.

185. R. J. Loy, Identities in tensor products of Banach algebras. Bull. Austral. Math. Soc. 2(1970), 253-260. MR 41 #5970. Z. 187, 384.

186. J. T. Marti, Introduction to the theory of bases. Springer-Verlag, New York, 1969. Z. 191, 413.

187. L. Maté, Multiplier operators and quotient algebra. Bull. Acad. Pol. Sci. 13(1965), 523-526. Z. 142, 6.

188. L. Maté, Embedding multiplier operators of a Banach algebra B into its second conjugate space B^{**}. Bull. Acad. Pol. Sci. 13(1965), 809-812. Z. 138, 382.

189. L. Maté, The Arens product and multiplier operators. Studia Math. 28(1967), 227-234. MR 35 #5938. Z. 158, 141.

190. C. W. McArthur, Developments in Schauder basis theory. Bull. Amer. Math. Soc. 78(1972), 877-908. MR 47 #2320. Z. 257.46012.

191. E. A. McCharen, A new topology on B^*-algebras arising from the Arens products. Proc. Amer. Math. Soc. 37(1973), 77-83. MR 46 #9736. Z. 252.46090.

BIBLIOGRAPHY

192. E. A. McCharen, A characterization of dual B*-algebras. Proc. Amer. Math. Soc. 37(1973), 84. MR 46 #6048. Z. 252.46091.

193. K. McKennon, Multipliers, positive functionals, positive-definite functions, and Fourier-Stieltjes transforms. Memoirs Amer. Math. Soc. 1971. Z. 214, 133.

194. K. McKennon, Quasi-multipliers. Trans. Amer. Math. Soc. 233(1977), 105-123. MR 56 #16376.

195. S. A. McKilligan, Duality in B*-algebras. Proc. Amer. Math. Soc. 38(1973), 86-88. MR 48 #4740. Z. 234.46061.

196. S. A. McKilligan, Arens regularity of certain Banach algebras which are ideals of a Banach algebra. Proc. Amer. Math. Soc. 50(1975), 223-229. MR 51 #8824. Z. 282.46048.

197. E. A. Michael, Locally multiplicatively-convex topological algebras. Memoirs Amer. Math. Soc. 11(1952). MR 14, 482. Z. 47, 355.

198. P. Milnes, Identities of group algebras. Proc. Amer. Math. Soc. 29(1971), 421-422. MR 43 #423. Z. 215, 406.

199. J. K. Miziolek, T. Müldner and A. Rek, On topologically nilpotent algebras. Studia Math. 43(1972), 41-50. MR 46 #6030. Z. 211, 159.

200. D. Montgomery and L. Zippin, Topological transformation groups. Interscience, New York, 1955. MR 17, 383.

201. R. D. Mosak, Central functions in group algebras. Proc. Amer. Soc. 29(1971), 613-616. MR 43 #5323. Z. 204, 353.

202. R. D. Mosak, The L^1-and C*-algebra of [FIA]$^-$ groups and their representations. Trans. Amer. Math. Soc. 163(1972), 277-310. MR 45 #2096. Z. 213, 136.

203. I. S. Murphy, Continuity of positive linear functionals on Banach *-algebras. Bull. London Math. Soc. 1(1969), 171-173. MR 40 #3320. Z. 181, 411.

204. M. A. Naimark, Normed algebras. Groningen, 1972. MR 22 #1824. Z. 254.46025.

205. T. Ogasawara, Finite-dimensionality of certain Banach algebras. J. Sci. Hiroshima Univ. Ser. A 17(1954), 359-364. MR 16, 598.

206. N. M. Osadčii, Approximate identities in certain Banach algebras of analytic functions. (Russian). Ukrain. Mat. Z. 28(1976), 408-412, 431. MR 54 #3419. Z. 341.46039.

BIBLIOGRAPHY

207. J. L. Ovaert, Factorisation dans les algèbres et modules de convolution. C. R. Acad. Sci. Paris 265(1967), 534-535. MR 36 #5714. Z. 175, 142.

208. W. L. Paschke, A factorable Banach algebra without bounded approximate unit. Pacific J. Math. 46(1973), 249-252. MR 48 #2765. Z. 257.46050.

209. A. Pelczyński, Any separable Banach space with the bounded approximation property is a complemented subspace of a Banach space with a basis. Studia Math. 40(1971), 239-243. MR 46 #7867. Z. 223.46019.

210. P. Porcelli, Linear spaces of analytic functions. Rand McNally & Co., Chicago, 1966. MR 41 #4219.

211. P. Porcelli and H. S. Collins, Ideals in group algebras. Studia Math. 33(1969), 223-226. MR 40 #4779. Z. 176, 442.

212. V. Pták, Un théorème de factorisation. C. R. Acad. Sci. Paris 275(1972), 1297-1299. MR 47 #831. Z. 252.46044.

213. V. Pták, Deux théorèmes de factorisation. C. R. Acad. Sci. Paris Sér. A 278(1974), 1091-1094. MR 49 #5846. Z. 277.46047.

214. V. Pták, A theorem of closed graph type. manuscripta math. 13(1974), 109-130. MR 50 #928. Z. 286.46008.

215. V. Pták, A quantitative refinement of the closed graph theorem. Czechoslovak Math J. 24(99) (1974), 503-506. MR 50 #929. Z. 315. 46007.

216. V. Pták, Factorization in Banach algebras. Studia Math. (to appear). Z. 342.46036.

217. T. Pytlik, On commutative approximate identities. Studia Math. 53(1975), 265-267. MR 51 #13586. Z. 319.43008.

218. H. Reiter, Classical harmonic analysis and locally compact groups. Oxford, 1968. MR 46 #5933. Z. 165, 156.

219. H. Reiter, Sur certains ideaux dans $L^1(G)$. C. R. Acad. Sci. Paris 267(1968), 882-885. MR 39 #6025. Z. 165, 158.

220. H. Reiter, L^1-algebras and Segal algebras. Lecture Notes in Mathematics 231. Springer-Verlag, New York, 1971. Z. 235.43004.

221. P. F. Renaud, Centralizers of the Fourier algebra of an amenable group. Proc. Amer. Math. Soc. 32(1972), 539-542. MR 45 #2099. Z. 258.43010.

BIBLIOGRAPHY

222. C. E. Rickart, General theory of Banach algebras. Van Nostrand, New York, 1960. MR 22 #5903. Z. 95, 97.

223. M. A. Rieffel, Induced Banach representations of Banach algebras and locally compact groups. J. Functional Analysis I (1967), 443-491. MR 36 #6544. Z. 181, 413.

224. M. A. Rieffel, On the continuity of certain intertwining operators, centralizers, and positive linear functionals. Proc. Amer. Math. Soc. 20(1969), 455-457. MR 38 #5000. Z. 177, 413.

225. M. A. Rieffel, Induced representations of C*-algebras. Advances in Mathematics 13(1974), 176-257. Z. 284.46040.

226. M. A. Rieffel, Morita equivalence for C*-algebras and W*-algebras. J. Pure and Applied Algebra 5(1974), 51-96. MR 51 #3912. Z. 295.46099.

227. M. A. Rieffel, Strong Morita equivalence of certain transformation group C*-algebras. Math. Ann. 222(1976), 7-22. MR 54 #7695. Z. 328.22013.

228. M. Riemersma, On some properties of normed ideals in $L^1(G)$. Nederl. Akad. Wetensch. Proc. Ser. A 78 = Indag. Math. 37(1975), 265-272. MR 52 #1170. Z. 304.43006.

229. H. Rindler, Approximierende Einheiten in Idealen von Gruppenalgebren. Anz. Österreich. Akad. Wiss. Math.-Naturwiss. Kl. 1976, no. 5, 37-39. MR 55 #565.

230. D. A. Robbins, Existence of a bounded approximate identity in a tensor product. Bull. Australian Math. Soc. 6(1972), 443-445. MR 46 #4231. Z. 229.46046.

231. A. P. Robertson and W. J. Robertson, Topological vector spaces. Cambridge Univ. Press, New York, 1964. MR 28 #5318. Z. 164, 430.

232. A. C. M. van Rooij and W. H. Schikhof, Non-archimedean commutative C*-algebras. Nederl. Akad. Wetensh. Proc., Ser. A 76 = Indag. Math 35(1973), 381-389. MR 48 #6963. Z. 272.46035.

233. H. P. Rosenthal, On the existence of approximate identities in ideals of group algebras. Arkiv Mat. 7(1967), 185-191. MR 39 #719. Z. 172, 183.

234. W. Rudin, Factorization in the group algebra of the real line. Proc. Nat. Acad. Sci. U. S. A. 43(1957), 339-340. MR 19, 46. Z. 79, 133.

BIBLIOGRAPHY

235. W. Rudin, Representation of functions by convolutions. J. Math. Mech. 7(1958), 103-115. MR 21 #3728. Z. 79, 133.

236. W. Rudin, Fourier analysis on groups. Interscience, New York, 1962. MR 27 #2808. Z. 107, 96.

237. W. Rudin, Spaces of type $H^\infty + C$. Ann. Inst. Fourier, Grenoble 25(1975), 99-125. MR 51 #13692. Z. 295.46080.

238. K. Saito, A characterization of double centralizer algebras of Banach algebras. Sci. Rep. Niigata Univ. Ser. A II(1974), 5-11. MR 49 #5844. Z. 349.46044.

239. S. Sakai, C^*-algebras and W^*-algebras. Springer-Verlag, New York, 1971. Z. 219, 292.

240. R. Salem, Sur les transformations des séries de Fourier. Fund. Math. 33(1939), 108-114. MR 1, 73.

241. R. Schatten, A theory of cross-spaces. Princeton University Press, Princeton, 1950. MR 12, 186. Z. 41, 435.

242. R. Schatten, Norm ideals of completely continuous operators. Springer-Verlag, New York, 1960. MR 22 #9878. Z. 90, 94.

243. I. E. Segal, Irreducible representations of operator algebras. Bull. Amer. Math. Soc. 53(1947), 73-88. MR 8, 520. Z. 31, 360.

244. I. E. Segal, The group algebra of a locally compact group. Trans. Amer. Math. Soc. 61(1947), 69-105. MR 8, 438. Z. 32, 29.

245. F. D. Sentilles and D. C. Taylor, Factorization in Banach algebras and the general strict topology. Trans. Amer. Math. Soc. 142(1969), 141-152. MR 40 #703. Z. 185, 211.

246. J. H. Shapiro, The bounded weak star topology and the general strict topology. J. Functional Analysis 8(1971), 275-286. MR 44 #7294. Z. 227.46023.

247. S. Shirali, Representability of positive functionals. J. London Math. Soc. 3(1971), 145-150. MR 43 #2517. Z. 236.46064.

248. J. E. Simpson, Locally convex algebras. III: Arens multiplication. Revue Roumaine Math. pur. appl. 19(1974), 359-370. Z. 277.46046.

249. A. M. Sinclair, Automatic continuity of linear operators. London Math. Soc. Lecture Note Series 21. Cambridge Univ. Press, Cambridge, 1976. Z. 313.47029.

250. A. M. Sinclair, Bounded approximate identities, factorization, and a convolution algebra. J. Functional Analysis 29(1978), 308-318. Z. 385.46030.

BIBLIOGRAPHY

251. A. M. Sinclair, Cohen's factorization method using an algebra of analytic functions. Preprint.

252. I. Singer, Bases in Banach spaces. I. Springer-Verlag, New York, 1970. MR 45 #7451. Z. 198, 166.

253. D. A. Stegenga, A note on spaces of type $H^{\infty} + C$. Ann. Inst. Fourier 25(1975), 213-217. MR 52 #11546. Z. 301.46041.

254. E. M. Stein, Singular integrals and differentiability properties of functions. Princeton Univ. Press, Princeton, N. J. 1970. MR 44 #7280. Z. 207, 135.

255. M. K. Summers, Factorization in Fréchet modules. J. London Math. Soc. 5(1972), 243-248. MR 47 #832. Z. 239.46043.

256. W. H. Summers, Factorization in Fréchet spaces. Studia Math. 39(1971), 209-216. MR 46 #6032. Z. 214, 137.

257. S. E. Takahashi, A remark on Archbolds result. Bull. Fac. Sci. Ibaraki Univ. Ser. A 8(1976), 23-24. MR 55 #1082. Z. 332.46039.

258. D. C. Taylor, A characterization of Banach algebras with approximate unit. Bull. Amer. Math. Soc. 74(1968), 761-766. MR 37 #2003. Z. 164, 446.

259. D. C. Taylor, A generalized Fatou theorem for Banach algebras. Pacific J. Math. 26(1968), 389-394. MR 38 #5007. Z. 165, 483.

260. D. C. Taylor, The strict topology for double centralizer algebras. Trans. Amer. Math. Soc. 150(1970), 633-643. MR 44 #7302. Z. 204, 147.

261. D. C. Taylor, A general Phillips theorem for C*-algebras and some applications. Pacific J. Math. 40(1972), 477-488. MR 46 #7913. Z. 239.46061.

262. D. C. Taylor, Interpolation in algebras of operator fields. J. Functional Analysis 10(1972), 159-190. MR 51 #13700. Z. 245.46083.

263. J. L. Taylor, Homology and cohomology for topological algebras. Advances in Mathematics 9(1972), 137-182. MR 48 #6967. Z. 271.46040.

264. U. B. Tewari, Multipliers of Segal algebras. Proc. Amer. Math. Soc. 54(1976), 157-161. Z. 317.43009.

265. B. J. Tomiuk and B. Yood, Topological algebras with dense socle. J. Functional Analysis, 28(1978), 254-277.

BIBLIOGRAPHY

266. J. Tomiyama, Primitive ideals in tensor products of Banach algebras. Math. Scand. 30(1972), 257-262. MR 48 #12079. Z. 245.46066.

267. C. N. Tseng and H. C. Wang, Closed subalgebras of homogeneous Banach algebras. J. Austral. Math. Soc. Ser.A 20(1975), 366-376. MR 54 #3298. Z. 309.43018.

268. K. R. Unni, A note on multipliers of Segal algebras. Studia Math. 49(1974), 125-127. MR 49 #5721. Z. 274.43008.

269. N. T. Varopoulos, Sur les formes positives d'une algèbre de Banach. C. R. Acad. Sci. Paris 258(1964), 2465-2467. Z. 131, 331.

270. N. T. Varopoulos, Sur les measures de Radon, d'un groupe localement compact abélien. C. R. Acad. Sci. Paris 258(1964), 3805-3808. MR 28 #4386. Z. 134, 122.

271. N. T. Varopoulos, Continuité des formes linéaires positives sur une algèbre de Banach avec involution. C. R. Acad. Sci. Paris 258(1964), 1121-1124. MR 28 #4387. Z. 148, 375.

272. J. von Neumann, Almost periodic functions in a group. I. Trans. Amer. Math. Soc. 36(1934), 445-492.

273. B. Walsh, Positive approximate identities and lattice-ordered dual spaces. Manuscripta math. 14(1974), 57-63. MR 50 #8000. Z. 305.46013.

274. H. C. Wang, Nonfactorization in group algebras. Studia Math. 42(1972), 231-241. MR 46 #2355. Z. 273.43008.

275. H. C. Wang, Factorization in homogeneous Banach algebras. Chinese J. Math. 1(1973), 199-202. MR 52 #6327.

276. H. C. Wang, Some results on homogeneous Banach algebras. Tamkang J. Math. 5(1974), 149-159. MR 53 #1172. Z. 313.43004.

277. H. C. Wang, Closed subalgebras of homogeneous Banach algebras. J. Austral. Math. Soc. Ser. A 20(1975), 366-376. MR 54 #3298. Z. 309.43018.

278. H. C. Wang, Homogeneous Banach algebras. Lecture Notes in Pure and Applied Math., 29, Marcel Dekker, New York, 1977. MR 56 #16258. Z. 346.43003.

279. C. R. Warner, Closed ideals in the group algebra $L^1(G) \cap L^2(G)$. Trans. Amer. Math. Soc. 121(1966), 408-423. MR 32 #4476. Z. 139, 306.

280. C. R. Warner and R. Whitley, A characterization of regular maximal ideals. Pacific J. Math. 30(1969), 277-281. Z. 176, 439.

BIBLIOGRAPHY

281. A. Weil, L'intégration dans les groupes topologiques et ses applications. Actualities Sci. et Ind. 869, 1145. Herman & Cie., Paris, 1941 and 1951.

282. H. Weyl and F. Peter, Die Vollständigkeit der primitiven Darstellungen einer geschlossenen kontinuierlichen Gruppe. Math. Ann. 97(1927), 737-755.

283. R. F. Wheeler, Well-behaved and totally bounded approximate identities for $C_o(X)$. Pacific J. Math. 65(1976), 261-269. MR 56 #16353. Z. 288.46049.

284. J. Wichmann, Bounded approximate units and bounded approximate identities. Proc. Amer. Math. Soc. 41(1973), 547-550. MR 48 #2767. Z. 272.46028.

285. D. Wildfogel, Double algebras, Segal algebras and spectral synthesis. Bull. Inst. Math., Acad. Sinica 2(1974), 241-254. MR 52 #6431. Z. 307.43016.

286. P. K. Wong, On the Arens product and annihilator algebras. Proc. Amer. Math. Soc. 30(1971), 79-83. MR 43 #6724. Z. 218, 295.

287. P. K. Wong, On the Arens products and certain Banach algebras. Trans. Amer. Math. Soc. 180(1973), 437-448. MR 47 #7431. Z. 238.46050.

288. L. Yap, Ideals in subalgebras of the group algebras. Studia Math. 35(1970), 165-175. MR 42 #4968. Z. 198, 178.

289. L. Yap, Every Segal algebra satisfies Ditkin's condition. Studia Math. 40(1971), 235-237. MR 46 #2358. Z. 226.43002.

290. L. Yap, On two classes of subalgebras of $L^1(G)$. Proc. Japan Acad. 48(1972), 315-319. MR 47 #717. Z. 253.43005.

291. L. Yap, Ideals in Fréchet subalgebras of $L^1(G)$. Math. Balkanica 3(1973), 610-613. MR 54 #13464. Z. 282.43006.

292. L. Yap, Nonfactorization of functions in Banach subspaces of $L^1(G)$. Proc. Amer. Math. Soc. 51(1975), 356-358. MR 53 #8782. Z. 309.43012.

293. Y. Zaïem, Opérateurs de convolution d'image fermée et unites approchées. Bull. Sci. Math. (2) 99(1975), 65-74. MR 55 #8690. Z. 317.43002.

SUBJECT INDEX

A-bimodule, 90

A-module, 90

abelian approximate identity, 2

abstract Segal algebra, 210

admissible norm, 46

algebra,
  amenable, 233
  Beurling, 179, 255
  C*- , 67
  Cohen factorable, 250
  disk, 235
  double centralizer, 196
  factorable, 250
  feebly factorable, 251
  Fréchet, 170
  function, 265
  group, 81
  group C*- , 88
  maximal commutative, 182
  quasicentral, 261
  quotient, 42
  reflexive, 225
  Segal, 262
  separable, 5
  symmetric *-, 242
  topological, 254
  von Neumann, 244
  weakly factorable, 250

algebraic tensor product, 45

amenable,
  algebra, 233
  group, 233, 246

approximate diagonal, 232

approximate identity, 2
  abelian, 2
  β-totally bounded, 258
  bound of, 2

approximate identity
  bounded, 2, 173
  canonical, 184
  central, 86
  chain totally bounded, 258
  commuting, 2
  countable, 2
  increasing, 69
  minimal, 265
  norm of, 2
  quasicentral, 197, 260
  right, 2
  semiaccurate, 40
  sequential, 2
  series, 244
  sham compact, 258
  two-sided, 2
  ultra- , 248
  uniform, 57
  uniformly bounded, 174
  weak, 223, 263
  well-behaved, 184

approximate units, 50
  bounded, 50
  left, 50
  pointwise bounded, 50
  right, 50
  two-sided, 50

approximation property, 204
  bounded, 205

analytic factorization, 140, 146

Arens product, 224
auxiliary Banach module, 115

β-topology, 181

β-totally bounded approximate identity, 258

Baire Category theorem, 56

SUBJECT INDEX

Banach *-algebraic bundle, 265

Banach module, 90
  auxiliary, 115
  essential, 91
  isometric, 91
  left, 90
  right, 90

basis, 204
  constant, 204
  shrinking, 207

Beurling algebra, 179, 255

bilinear mapping, 226

bimodule, 90

bound of approximate identity, 2

bounded approximate identity, 2, 173

bounded approximate units, 50

bounded approximation property, 205

bounded multiple units, 237

Buck topology, 255

C^*-algebra, 67

canonical approximate identity, 184

central approximate identity, 86

centralizer, 119, 196

chain totally bounded approximate identity, 258

closed graph theorem, 100

cohomology group, 234

coefficient functional, 204

Cohen element, 253

Cohen factorable, 250

compact operator, 203

completely positive lifting, 260

contractor, 166

converse of fundamental theorem, 105

cross-norm, 45
  greatest, 45
  least, 46

cover,
  decomposable, 190
  equivalent, 190
  good, 192
  well-behaved, 189

derivation, 234
  inner, 234

double centralizer, 196

element,
  positive, 69
  strictly positive, 72

equivalent cover, 190

essential Banach module, 91

essential Fréchet module, 172

essential part, 92, 172

extreme point, 256, 257

extremally disconnected, 191

factorable, 250
  Cohen, 250
  feebly, 251
  weakly, 250

SUBJECT INDEX

factorable algebra without
  approximate units, 161

factorization, 93
  analytic, 140, 146
  multiple, 114
  power, 122

feebly factorable, 251

first cohomology group, 234

Fréchet algebra, 170

Fréchet module, 170
  essential, 172
  left, 170
  right, 170

function algebra, 265

fundamental factorization
  theorem, 93

Gelfand-Naimark theorem, 67, 69, 241

good cover, 192

good function, 192

global property, 182

greatest cross-norm, 45

group algebra, 81

group C^*-algebra, 88

Hilbert-Schmidt operators, 262

increasing approximate
  identity, 69

inner derivation, 234

intermediate value theorem, 15

invariant mean, 235

isometric Banach module, 91

least cross-norm, 46

left approximate identity, 2

left approximate units, 50

local property, 182

majorization theorem, 229

matrix ranges, 260

maximal commutative
  *-subalgebra, 182

method of linear functionals, 219

multiple factorization, 114

multiple units, 237

nonfactorization, 167

nonsequential approximate identities,
  33

norm of an approximate identity, 2

normal element, 182

normed module, 90

paracompact, 186, 188, 242

perfect map, 188

pointwise-bounded approximate units, 50

SUBJECT INDEX

positive element, 69

positive functional, 69, 119

projection, 204

pseudocompact, 257

pure state, 243

quasicentral approximate identity, 197, 260

quasicentral C^*-algebra, 261

quotient algebra, 42

reflexive Banach algebra, 225

regular Arens product, 225

renorming approximate identities, 17

right approximate identity, 2

right approximate units, 50

second dual space, 224, 244

Segal algebra, 262

selfadjoint element, 182

semiaccurate approximate identity, 40

semigroup, 5, 19, 129

semigroup algebra, 166, 263

separable algebra, 5

sequential approximate identity, 2

series approximate identity, 244

sham compact approximate identity, 258

sham compact topological space, 258

shrinking basis, 207

SIN-group, 86, 244, 246

strict topology, 181, 255

strictly positive element, 72

sum of subspaces, 221

summability kernel, 237

symmetric *-algebra, 242

tensor product, 44

topological algebra, 254

topological zero divisor, 57

topologically nilpotent element, 60

topologically nilpotent normed algebra, 63

totally bounded subset, 116

trace class operators, 262

two-sided approximate identity, 2

two-sided approximate units, 50

unbounded approximate identities, 63

uniform boundedness theorem, 13

SUBJECT INDEX

uniform approximate identity, 57

uniformly bounded approximate identity, 174

weak approximate identity, 223, 263

weakly factorable algebra, 250

well-behaved approximate identity, 184

well-behaved cover, 189

word, 6

zero-dimensional space, 188

SYMBOL INDEX

$(ab)^{\#}$, 164

A_1, 130

$(A,||.||_A)$, 210

$A^p(G)$, 167

A_n, 179

A_r, 225

$\tilde{A}$, 57

A^*, 48

$A^* = A\cdot A^*$, 198

$A \otimes B$, 45

A/I, 42

AX, 92

B'(X), 203

B(H), 258

$B^1(A,X)$, 234

$B(x,\delta)$, 56

c_o, 146

C_β, 36

C^2, 79

C(E), 35

$C_r^*(G)$, 87

$C^*(G)$, 88

$C_b(S)$, 258

C_p, 262

$C_{oo}(G)$, 81

$C_o(G)$, 67

$D(\beta)$, 135

$\Delta(\beta)$, 135

$\Delta(\cdot)$, 82

δ_x, 234

e_n^*, 204

exp x, 130

$E \otimes^\alpha F$, 45

$E \otimes_\alpha F$, 45

$f \otimes g$, 196

$f \to ||T_f||$, 87

(f,g), 84

${}_af$, 82

f_a, 82

f*g, 81

$f^*(x)$, 82

$\langle f,a \rangle$, 219

F_α, 148

[FD], 246

$[FD]^-$, 246

$[FC]^-$, 246

$[FIA]^-$, 246

FG, 224

$G(\beta)$, 149

h_m, 167

H, 150

$\underline{H}$, 24

H_m, 167

$H^1(A,X)$, 234

$H^r(A)$, 57

$\underline{H}'(n)$, 31

H^∞, 263

Hol(U), 147

I, 48

J_b, 121

K(x), 50

SYMBOL INDEX

ℓ^p, 113

$\ell^1(S)$, 166

$\ell^\infty(G)$, 235

l.a.i., 2

l.a.u., 50

$L(0,1)$, 61

L^∞, 263

$L(X)$, 41

$L^1(G)$, 81

$L^2(G)$, 86

$L^o(G)$, 245

$M(a)$, 235

$M(A)$, 196

$M(A)_\beta$, 258

$M(G)$, 245

$M^o(G)$, 245

$M^\perp$, 72

$N(A)$, 61

$O(x)$, 41

$\theta(x)$, 35

$\mathcal{O}$, 253

P_i, 204

r.a.i., 2

r.a.u., 50

R^+, 24

$R(a_n,\alpha_i)$, 143

(R,S), 196

$(R,S)^*$, 197

$s(E,x)$, 25

$S(E,x)$, 25

$s(\beta_1,\ldots,\beta_n,x)$, 37

$S(C,f_{n\varepsilon})$, 26

[SIN], 86, 246

$\sigma(r)$, 100

$\sigma(a_n)$, 141

$\sigma(A^{**},A^*)$, 244

t.a.i., 2

t.a.u., 50

T_a, 49

T_a^*, 49

$T_g h$, 235

$T_n x$, 13

$w(b)$, 167

$W(r)$, 100

$\omega_o(x)$, 179

ω_1, 39

ω_2, 34

$\omega^n(.)$, 100

X_e, 92

$X \times Y$, 195

$x \otimes y$, 45

$x \leq y$, 69

$x \geq 0$, 69

$Y + Z$, 221

Z^+, 24

$Z(A)$, 229

$Z^1(A,X)$, 234

$Z^r(A)$, 57

π, 232

π_{ij}, 170

$\tilde{\pi}_{ij}$, 170

π_p, 75

Γ_n, 80

$\Gamma(n)$, 148

Vol. 609: General Topology and Its Relations to Modern Analysis and Algebra IV. Proceedings 1976. Edited by J. Novák. XVIII, 225 pages. 1977.

Vol. 610: G. Jensen, Higher Order Contact of Submanifolds of Homogeneous Spaces. XII, 154 pages. 1977.

Vol. 611: M. Makkai and G. E. Reyes, First Order Categorical Logic. VIII, 301 pages. 1977.

Vol. 612: E. M. Kleinberg, Infinitary Combinatorics and the Axiom of Determinateness. VIII, 150 pages. 1977.

Vol. 613: E. Behrends et al., L^p-Structure in Real Banach Spaces. X, 108 pages. 1977.

Vol. 614: H. Yanagihara, Theory of Hopf Algebras Attached to Group Schemes. VIII, 308 pages. 1977.

Vol. 615: Turbulence Seminar, Proceedings 1976/77. Edited by P. Bernard and T. Ratiu. VI, 155 pages. 1977.

Vol. 616: Abelian Group Theory, 2nd New Mexico State University Conference, 1976. Proceedings. Edited by D. Arnold, R. Hunter and E. Walker. X, 423 pages. 1977.

Vol. 617: K. J. Devlin, The Axiom of Constructibility: A Guide for the Mathematician. VIII, 96 pages. 1977.

Vol. 618: I. I. Hirschman, Jr. and D. E. Hughes, Extreme Eigen Values of Toeplitz Operators. VI, 145 pages. 1977.

Vol. 619: Set Theory and Hierarchy Theory V, Bierutowice 1976. Edited by A. Lachlan, M. Srebrny, and A. Zarach. VIII, 358 pages. 1977.

Vol. 620: H. Popp, Moduli Theory and Classification Theory of Algebraic Varieties. VIII, 189 pages. 1977.

Vol. 621: Kauffman et al., The Deficiency Index Problem. VI, 112 pages. 1977.

Vol. 622: Combinatorial Mathematics V, Melbourne 1976. Proceedings. Edited by C. Little. VIII, 213 pages. 1977.

Vol. 623: I. Erdelyi and R. Lange, Spectral Decompositions on Banach Spaces. VIII, 122 pages. 1977.

Vol. 624: Y. Guivarc'h et al., Marches Aléatoires sur les Groupes de Lie. VIII, 292 pages. 1977.

Vol. 625: J. P. Alexander et al., Odd Order Group Actions and Witt Classification of Innerproducts. IV, 202 pages. 1977.

Vol. 626: Number Theory Day, New York 1976. Proceedings. Edited by M. B. Nathanson. VI, 241 pages. 1977.

Vol. 627: Modular Functions of One Variable VI, Bonn 1976. Proceedings. Edited by J.-P. Serre and D. B. Zagier. VI, 339 pages. 1977.

Vol. 628: H. J. Baues, Obstruction Theory on the Homotopy Classification of Maps. XII, 387 pages. 1977.

Vol. 629: W.A. Coppel, Dichotomies in Stability Theory. VI, 98 pages. 1978.

Vol. 630: Numerical Analysis, Proceedings, Biennial Conference, Dundee 1977. Edited by G. A. Watson. XII, 199 pages. 1978.

Vol. 631: Numerical Treatment of Differential Equations. Proceedings 1976. Edited by R. Bulirsch, R. D. Grigorieff, and J. Schröder. X, 219 pages. 1978.

Vol. 632: J.-F. Boutot, Schéma de Picard Local. X, 165 pages. 1978.

Vol. 633: N. R. Coleff and M. E. Herrera, Les Courants Résiduels Associés à une Forme Méromorphe. X, 211 pages. 1978.

Vol. 634: H. Kurke et al., Die Approximationseigenschaft lokaler Ringe. IV, 204 Seiten. 1978.

Vol. 635: T. Y. Lam, Serre's Conjecture. XVI, 227 pages. 1978.

Vol. 636: Journées de Statistique des Processus Stochastiques, Grenoble 1977, Proceedings. Edité par Didier Dacunha-Castelle et Bernard Van Cutsem. VII, 202 pages. 1978.

Vol. 637: W. B. Jurkat, Meromorphe Differentialgleichungen. VII, 194 Seiten. 1978.

Vol. 638: P. Shanahan, The Atiyah-Singer Index Theorem, An Introduction. V, 224 pages. 1978.

Vol. 639: N. Adasch et al., Topological Vector Spaces. V, 125 pages. 1978.

Vol. 640: J. L. Dupont, Curvature and Characteristic Classes. X, 175 pages. 1978.

Vol. 641: Séminaire d'Algèbre Paul Dubreil, Proceedings Paris 1976–1977. Edité par M. P. Malliavin. IV, 367 pages. 1978.

Vol. 642: Theory and Applications of Graphs, Proceedings, Michigan 1976. Edited by Y. Alavi and D. R. Lick. XIV, 635 pages. 1978.

Vol. 643: M. Davis, Multiaxial Actions on Manifolds. VI, 141 pages. 1978.

Vol. 644: Vector Space Measures and Applications I, Proceedings 1977. Edited by R. M. Aron and S. Dineen. VIII, 451 pages. 1978.

Vol. 645: Vector Space Measures and Applications II, Proceedings 1977. Edited by R. M. Aron and S. Dineen. VIII, 218 pages. 1978.

Vol. 646: O. Tammi, Extremum Problems for Bounded Univalent Functions. VIII, 313 pages. 1978.

Vol. 647: L. J. Ratliff, Jr., Chain Conjectures in Ring Theory. VIII, 133 pages. 1978.

Vol. 648: Nonlinear Partial Differential Equations and Applications, Proceedings, Indiana 1976–1977. Edited by J. M. Chadam. VI, 206 pages. 1978.

Vol. 649: Séminaire de Probabilités XII, Proceedings, Strasbourg, 1976–1977. Edité par C. Dellacherie, P. A. Meyer et M. Weil. VIII, 805 pages. 1978.

Vol. 650: C*-Algebras and Applications to Physics. Proceedings 1977. Edited by H. Araki and R. V. Kadison. V, 192 pages. 1978.

Vol. 651: P. W. Michor, Functors and Categories of Banach Spaces. VI, 99 pages. 1978.

Vol. 652: Differential Topology, Foliations and Gelfand-Fuks-Cohomology, Proceedings 1976. Edited by P. A. Schweitzer. XIV, 252 pages. 1978.

Vol. 653: Locally Interacting Systems and Their Application in Biology. Proceedings, 1976. Edited by R. L. Dobrushin, V. I. Kryukov and A. L. Toom. XI, 202 pages. 1978.

Vol. 654: J. P. Buhler, Icosahedral Golois Representations. III, 143 pages. 1978.

Vol. 655: R. Baeza, Quadratic Forms Over Semilocal Rings. VI, 199 pages. 1978.

Vol. 656: Probability Theory on Vector Spaces. Proceedings, 1977. Edited by A. Weron. VIII, 274 pages. 1978.

Vol. 657: Geometric Applications of Homotopy Theory I, Proceedings 1977. Edited by M. G. Barratt and M. E. Mahowald. VIII, 459 pages. 1978.

Vol. 658: Geometric Applications of Homotopy Theory II, Proceedings 1977. Edited by M. G. Barratt and M. E. Mahowald. VIII, 487 pages. 1978.

Vol. 659: Bruckner, Differentiation of Real Functions. X, 247 pages. 1978.

Vol. 660: Equations aux Dérivée Partielles. Proceedings, 1977. Edité par Pham The Lai. VI, 216 pages. 1978.

Vol. 661: P. T. Johnstone, R. Paré, R. D. Rosebrugh, D. Schumacher, R. J. Wood, and G. C. Wraith, Indexed Categories and Their Applications. VII, 260 pages. 1978.

Vol. 662: Akin, The Metric Theory of Banach Manifolds. XIX, 306 pages. 1978.

Vol. 663: J. F. Berglund, H. D. Junghenn, P. Milnes, Compact Right Topological Semigroups and Generalizations of Almost Periodicity. X, 243 pages. 1978.

Vol. 664: Algebraic and Geometric Topology, Proceedings, 1977. Edited by K. C. Millett. XI, 240 pages. 1978.

Vol. 665: Journées d'Analyse Non Linéaire. Proceedings, 1977. Edité par P. Bénilan et J. Robert. VIII, 256 pages. 1978.

Vol. 666: B. Beauzamy, Espaces d'Interpolation Réels: Topologie et Géometrie. X, 104 pages. 1978.

Vol. 667: J. Gilewicz, Approximants de Padé. XIV, 511 pages. 1978.

Vol. 668: The Structure of Attractors in Dynamical Systems. Proceedings, 1977. Edited by J. C. Martin, N. G. Markley and W. Perrizo. VI, 264 pages. 1978.

Vol. 669: Higher Set Theory. Proceedings, 1977. Edited by G. H. Müller and D. S. Scott. XII, 476 pages. 1978.

Vol. 670: Fonctions de Plusieurs Variables Complexes III, Proceedings, 1977. Edité par F. Norguet. XII, 394 pages. 1978.

Vol. 671: R. T. Smythe and J. C. Wierman, First-Passage Perculation on the Square Lattice. VIII, 196 pages. 1978.

Vol. 672: R. L. Taylor, Stochastic Convergence of Weighted Sums of Random Elements in Linear Spaces. VII, 216 pages. 1978.

Vol. 673: Algebraic Topology, Proceedings 1977. Edited by P. Hoffman, R. Piccinini and D. Sjerve. VI, 278 pages. 1978.

Vol. 674: Z. Fiedorowicz and S. Priddy, Homology of Classical Groups Over Finite Fields and Their Associated Infinite Loop Spaces. VI, 434 pages. 1978.

Vol. 675: J. Galambos and S. Kotz, Characterizations of Probability Distributions. VIII, 169 pages. 1978.

Vol. 676: Differential Geometrical Methods in Mathematical Physics II, Proceedings, 1977. Edited by K. Bleuler, H. R. Petry and A. Reetz. VI, 626 pages. 1978.

Vol. 677: Séminaire Bourbaki, vol. 1976/77, Exposés 489–506. IV, 264 pages. 1978.

Vol. 678: D. Dacunha-Castelle, H. Heyer et B. Roynette. Ecole d'Eté de Probabilités de Saint-Flour. VII-1977. Edité par P. L. Hennequin. IX, 379 pages. 1978.

Vol. 679: Numerical Treatment of Differential Equations in Applications, Proceedings, 1977. Edited by R. Ansorge and W. Törnig. IX, 163 pages. 1978.

Vol. 680: Mathematical Control Theory, Proceedings, 1977. Edited by W. A. Coppel. IX, 257 pages. 1978.

Vol. 681: Séminaire de Théorie du Potentiel Paris, No. 3, Directeurs: M. Brelot, G. Choquet et J. Deny. Rédacteurs: F. Hirsch et G. Mokobodzki. VII, 294 pages. 1978.

Vol. 682: G. D. James, The Representation Theory of the Symmetric Groups. V, 156 pages. 1978.

Vol. 683: Variétés Analytiques Compactes, Proceedings, 1977. Edité par Y. Hervier et A. Hirschowitz. V, 248 pages. 1978.

Vol. 684: E. E. Rosinger, Distributions and Nonlinear Partial Differential Equations. XI, 146 pages. 1978.

Vol. 685: Knot Theory, Proceedings, 1977. Edited by J. C. Hausmann. VII, 311 pages. 1978.

Vol. 686: Combinatorial Mathematics, Proceedings, 1977. Edited by D. A. Holton and J. Seberry. IX, 353 pages. 1978.

Vol. 687: Algebraic Geometry, Proceedings, 1977. Edited by L. D. Olson. V, 244 pages. 1978.

Vol. 688: J. Dydak and J. Segal, Shape Theory. VI, 150 pages. 1978.

Vol. 689: Cabal Seminar 76–77, Proceedings, 1976–77. Edited by A.S. Kechris and Y. N. Moschovakis. V, 282 pages. 1978.

Vol. 690: W. J. J. Rey, Robust Statistical Methods. VI, 128 pages. 1978.

Vol. 691: G. Viennot, Algèbres de Lie Libres et Monoïdes Libres. III, 124 pages. 1978.

Vol. 692: T. Husain and S. M. Khaleelulla, Barrelledness in Topological and Ordered Vector Spaces. IX, 258 pages. 1978.

Vol. 693: Hilbert Space Operators, Proceedings, 1977. Edited by J. M. Bachar Jr. and D. W. Hadwin. VIII, 184 pages. 1978.

Vol. 694: Séminaire Pierre Lelong – Henri Skoda (Analyse) Année 1976/77. VII, 334 pages. 1978.

Vol. 695: Measure Theory Applications to Stochastic Analysis, Proceedings, 1977. Edited by G. Kallianpur and D. Kölzow. XII, 261 pages. 1978.

Vol. 696: P. J. Feinsilver, Special Functions, Probability Semigroups, and Hamiltonian Flows. VI, 112 pages. 1978.

Vol. 697: Topics in Algebra, Proceedings, 1978. Edited by M. F. Newman. XI, 229 pages. 1978.

Vol. 698: E. Grosswald, Bessel Polynomials. XIV, 182 pages. 1978.

Vol. 699: R. E. Greene and H.-H. Wu, Function Theory on Manifolds Which Possess a Pole. III, 215 pages. 1979.

Vol. 700: Module Theory, Proceedings, 1977. Edited by C. Faith and S. Wiegand. X, 239 pages. 1979.

Vol. 701: Functional Analysis Methods in Numerical Analysis, Proceedings, 1977. Edited by M. Zuhair Nashed. VII, 333 pages. 1979.

Vol. 702: Yuri N. Bibikov, Local Theory of Nonlinear Analytic Ordinary Differential Equations. IX, 147 pages. 1979.

Vol. 703: Equadiff IV, Proceedings, 1977. Edited by J. Fábera. XIX, 441 pages. 1979.

Vol. 704: Computing Methods in Applied Sciences and Engineering, 1977, I. Proceedings, 1977. Edited by R. Glowinski and J. L. Lions. VI, 391 pages. 1979.

Vol. 705: O. Forster und K. Knorr, Konstruktion verseller Familien kompakter komplexer Räume. VII, 141 Seiten. 1979.

Vol. 706: Probability Measures on Groups, Proceedings, 1978. Edited by H. Heyer. XIII, 348 pages. 1979.

Vol. 707: R. Zielke, Discontinuous Čebyšev Systems. VI, 111 pages. 1979.

Vol. 708: J. P. Jouanolou, Equations de Pfaff algébriques. V, 255 pages. 1979.

Vol. 709: Probability in Banach Spaces II. Proceedings, 1978. Edited by A. Beck. V, 205 pages. 1979.

Vol. 710: Séminaire Bourbaki vol. 1977/78, Exposés 507–524. IV, 328 pages. 1979.

Vol. 711: Asymptotic Analysis. Edited by F. Verhulst. V, 240 pages. 1979.

Vol. 712: Equations Différentielles et Systèmes de Pfaff dans le Champ Complexe. Edité par R. Gérard et J.-P. Ramis. V, 364 pages. 1979.

Vol. 713: Séminaire de Théorie du Potentiel, Paris No. 4. Edité par F. Hirsch et G. Mokobodzki. VII, 281 pages. 1979.

Vol. 714: J. Jacod, Calcul Stochastique et Problèmes de Martingales. X, 539 pages. 1979.

Vol. 715: Inder Bir S. Passi, Group Rings and Their Augmentation Ideals. VI, 137 pages. 1979.

Vol. 716: M. A. Scheunert, The Theory of Lie Superalgebras. X, 271 pages. 1979.

Vol. 717: Grosser, Bidualräume und Vervollständigungen von Banachmoduln. III, 209 pages. 1979.

Vol. 718: J. Ferrante and C. W. Rackoff, The Computational Complexity of Logical Theories. X, 243 pages. 1979.

Vol. 719: Categorial Topology, Proceedings, 1978. Edited by H. Herrlich and G. Preuß. XII, 420 pages. 1979.

Vol. 720: E. Dubinsky, The Structure of Nuclear Fréchet Spaces. V, 187 pages. 1979.

Vol. 721: Séminaire de Probabilités XIII. Proceedings, Strasbourg, 1977/78. Edité par C. Dellacherie, P. A. Meyer et M. Weil. VII, 647 pages. 1979.

Vol. 722: Topology of Low-Dimensional Manifolds. Proceedings, 1977. Edited by R. Fenn. VI, 154 pages. 1979.

Vol. 723: W. Brandal, Commutative Rings whose Finitely Generated Modules Decompose. II, 116 pages. 1979.

Vol. 724: D. Griffeath, Additive and Cancellative Interacting Particle Systems. V, 108 pages. 1979.

Vol. 725: Algèbres d'Opérateurs. Proceedings, 1978. Edité par P. de la Harpe. VII, 309 pages. 1979.

Vol. 726: Y.-C. Wong, Schwartz Spaces, Nuclear Spaces and Tensor Products. VI, 418 pages. 1979.

Vol. 727: Y. Saito, Spectral Representations for Schrödinger Operators With Long-Range Potentials. V, 149 pages. 1979.

Vol. 728: Non-Commutative Harmonic Analysis. Proceedings, 1978. Edited by J. Carmona and M. Vergne. V, 244 pages. 1979.